Improving Health Care

Case Studies from Leading Practitioners

A Journal of Innovative Management Collection; formerly
The Business of Health Care

Editor: Laurence R. Smith
Designer: Janet MacCausland

GOAL/QPC

12B Manor Parkway
Salem, NH 03079
603.893.1944
www.goalqpc.com

Table of Contents

Improving Health Care

Health care has safety and quality problems because it relies on outmoded systems of work. Poor designs set the workforce up to fail, regardless of how hard they try. If we want safer, higher-quality care, we will need to have redesigned systems of care.... — Crossing the Quality Chasm, I.O.M.

High quality health care is more than accurate diagnosis and treatment decisions. It includes the quality of attitudes and relationships of everyone in the system. It includes the quality of the processes and systems for the delivery of care. It includes the quality of management and the actual delivery of care in a complex and sometimes stressful environment.

In a very important sense health and quality—personal and organizational—are about how we want to live and work, and how to design and manage effective processes and systems to consistently deliver the care prescribed.

Regardless of our profession or occupation in life, we all want to have a good quality of life, and that generally includes as much health as possible. Clearly when we are sick or injured we want caregivers, processes, and systems that are optimized to treat the disease or injury; but those are merely paths to restored health. We tolerate sick care processes when necessary but we simultaneously want health care to achieve improved quality of life.

There are good examples where this is being done. The articles in this book, first published in the *Journal of Innovative Management*, showcase some of these examples. They are assembled in this single volume for easy access.

Improving Patient Care in Hospitals, by Paul Uhlig, M.D., is about an impressive cardiac services caregivers team at Concord Hospital, Concord, New Hampshire. The program is called the Concord Care Collaborative Model and it was recently awarded the John M. Eisenberg Patient Safety Award by the National Quality Forum and the Joint Commission on Accreditation of Healthcare Organizations.

In Six Sigma in Health Care: A Road Less Traveled, Joseph Calvaruso, President & CEO of Mt. Carmel Health Systems in Columbus, Ohio, showcases their Six Sigma start up. Calvaruso says they experienced a 70% increase in net operating income, a 144% increase in net income, and a 8.7% increase in patient revenues.

These achievements come from a combination of Six Sigma and their new leadership development program. It includes improving processes so that people enjoy the work they're doing more, and caregivers can spend more time giving care and being with patients.

In Process Design and Management: The Path to Organizational Transformation, Sr. Mary Jean Ryan, FSM, President and CEO of SSM Health Care in St. Louis, tells us that something powerful happens in an organization when people work on process improvement. Shey says, "Looking for ways to improve the way we do things for our customers and clients alters the vantage point from which we see. And by altering our perspective, we transform what we have been dealing with–even if we have been dealing with it for years."

In Living and Breathing a Customer-Centered Culture, by Christine Kelly, Director of Laboratory, Health System Minnesota, and Elizabeth Lentz, Regional Laboratory Manager, Park Nicollett Clinic, Health System Minnesota, Minneapolis, Minnesota, the authors tell us that in 1993 the organization started to move away from a traditional approach to quality improvement to one of transformation. "In a transformed organization there is continuous learning about quality improvement through application in daily work processes rather than separate, disconnected, training sessions. There is global understanding of the organization's systems and processes, as well as the capabilities of the current systems."

The authors of Putting the Patient at the Core of a Healthcare Organization, Tom Tibbitts, President, and Sue Thompson, VP, Patient Support Services, Trinity Health Systems, Ft. Dodge, Iowa, discuss their restructuring and reengineering processes, and show how it has improved their ability to provide a more efficient, patient-centered type of care. Financial savings, plus positive results in patient and physician satisfaction, have been achieved.

In From Incremental to Breakthrough Performance, Ellen J. Gaucher, Senior Assoociate Director and COO, University of Michigan Hospitals, Ann Arbor, Michigan, tells us that breakthrough performance begins with the individual. Leaders must step above personal interest and consider what is best for the long term of their organization. Leaders must also commit personal energy to create a sense of urgency, craft and communicate the vision, set stretch goals and expectations for change, personally focus on customers, keep the pressure on, and demonstrate continuous commitment through actions and decisions.

In Hoshin Planning in Health Care, by Geoffrey Crabtree, VP, Strategic Planning & Market Services, Methodist Healthcare System, San Antonio, Texas; Dona Hotopp, Director of Healthcare Services, GOAL/QPC, Salem, New Hampshire; and Owen McNally, Director of Total Quality Management, Our Lady of Lourdes Medical Center, Camden, New Jersey.

Hoshin is a planning system to diagnose processes, compare actual and target results, analyze the cause of any gaps, and tie it all to corporate vision. It was developed in Japan in the mid 1960s and introduced in health care in 1990. Our Lady of Lourdes started the Hoshin process with its strategic plan, while Methodist Healthcare Systems used Hoshin to develop their customer-focused strategic plan. The authors say it has brought strategic plans to life, and while cost cutting was not an objective, cost reduction and revenue growth happened as a result.

In Managing by Teams at Rush Home Care Network, Kathryn E. Christiansen, D.N.Sc., Executive Director of Rush Home Care Network, Rush-Presbyterian-St. Luke's Medical Center in Chicago, Illinois, describes the hopes, successes and difficulties of changing a system and culture to create interdisciplinary care teams. She says, "Everyone must know and understand that everything they do affects someone else (internal or external to the organization) and contributes to our success or lack of success. The all-one-team concept is more than words. It really means that teams are more successful than individuals alone. It also means that one team cannot become insular in their outlook."

In From a Culture of Safety to a Culture of Excellence, Martin D. Merry, M.D. C.M., Senior Medical Advisor to the New Hampshire Hospital Association and Associate Professor of Health Management at the University of New Hampshire in Exeter, New Hampshire, and Jeffrey P. Brown, M.Ed., Principal of the System Safety Group in Peterborough, New Hampshire, share a wealth of process and systems experience. The authors say: "The healthcare (non)system on which we rely in our most vulnerable moments has in recent decades grown (1) far too large and complex for the craft model on which it was built and (2) far more dangerous than anyone, until recently, has realized. Because it has no other real choice, health care will embrace management systems guru W. Edwards Deming's dictum that improving quality through process improvement simultaneously decreases cost and enhances value delivered."

In Process Management and Systems Thinking for Patient Safety by Joanne E. Turnbull, Ph.D., Executive Director, National Patient Safety Foundation, Chicago, Illinois. We are provided an insightful understanding of the problem and what it will take to fix it. Turnbull lays out an eight-level systems change program called the "Onion Model." It includes understanding and harnessing complexity, aligning all the agents that are acting in the system, and reconnecting to the core business.

In Untangling the Web: Bringing Information Therapy to the New Healthcare Consumer, Molly Mettler, Sr. VP of Healthwise, Inc. in Boise, Idaho, says that Healthwise believes that to have a better healthcare system, the role of the patient must first be reinvented. Their vision is to build a better patient by creating communities of the best informed, most empowered healthcare consumers in the world.

In Chaos Theory and Creativity: The Biological Basis of Innovation, Ary Goldberger, M.D., Associate Professor of Medicine at Harvard Medical School and Director of the Electrocardiography and Arrhythmia Monitoring Laboratory at Beth Israel Deaconness Medical Center in Boston, Massachusetts, says there are three defining characteristics of a healthy individual and organization: (1) Productivity–the ability to do useful things. (2) Innovation–the ability to grow and change. (3) Resilience–the ability to bounce back from an injury or setback; the ability to heal.

In The Future of Medicine, Andrew Weil, M.D., Director of the Integrative Medicine Program and Clinical Professor of Medicine at the University of Arizona Medical Center in Tuscon, Arizona, says: "Conventional medicine has become too expensive and is threatening to sink our healthcare system. By enlarging and expanding the paradigms of conventional medicine, practitioners have the opportunity to become better healers and to better serve their patients while revitalizing the healthcare industry."

In Counteracting the Harmful Effects of Stress through Self-Care to Enhance Wellness and Profitability, Herbert Benson, M.D., President of the Mind/Body Medical Institute and Associate Professor of Medicine at Harvard Medical School, Boston, Massachusetts, says it is estimated that job stress costs employers $2 billion annually in absenteeism, sub-par performance, tardiness, and worker's compensation claims. Research has shown that more than 50% of adults report high stress every day and that stress is directly linked to numerous medical conditions, such as hypertension, asthma, chronic pain, and allergies, which can account for significant job absenteeism. Although mind/body therapies have been proven effective for the vast majority of everyday medical problems, we are still far more likely to run to our medicine cabinet to relieve our aches and pains than to consider using relaxation or stress-management techniques.

Improving Patient Care in Hospitals

Author

Paul N. Uhlig, M.D., Medical Director of the Cardiac Surgery Service, Concord Hospital, Concord, New Hampshire; Associate Professor of Surgery, Dartmouth Medical School, Hanover, New Hampshire

Introduction

In the early 1990s things began to change in my profession. It is fair to say that my heart began to break for some of the changes that I was seeing in health care. From my perspective, it seemed that many of the fundamental values I had been taught, literally at my father's knee, weren't being built into the systems of care being created. I didn't understand why these particular changes were happening, and I felt angry that dollars were being put between patients and doctors. I wanted to understand what was happening, which led me through a course of self-study and, five years ago, back to school for a year at Harvard where I received a Master of Public Administration degree in healthcare policy. During that time I had the opportunity to reflect about the changes occurring in health care—why they are happening, what they mean for us as practitioners, and what they mean for patients.

After my year at Harvard and several years in Kansas, I was offered a position at Dartmouth Medical School, which I accepted as an opportunity to explore some ideas that I had been formulating for some time about how doctors, nurses, patients, and others can work together in better ways. I'm happy to report some of that work to you in this article. I believe that everyone reading this—those of us in health care and those in other fields—will be able to appreciate this because all of us, in our own ways, are pioneers, explorers of ways of thinking and acting that we believe have promise for the future for our respective professions.

What does "health care" mean?

When we use the term "health care," just what do we mean? Where does it all begin? I believe health care began when the first primordial creature came stumbling out of the swamp and the creature next to it reached down and tried to help drag it out. I think this is the fundamental idea operating in health care. That is, we help each other because we never know when we will need that same help ourselves. Health care emerges from relationships founded on trust and, ultimately, in shared human experience. A sense of interdependence and reciprocity underlies what we do in health care, and in life. This connection with some of the most fundamental aspects of our being is what makes work in health care so special, and our responsibilities so great.

The challenge of transforming health care

This is a time of profound transformation in health care. The decisions we are making, and the actions we are taking today, will determine the care that we and those we love will receive in the years ahead. We must take this work very seriously. We face important challenges. Despite our best efforts, costs are very high, quality is uneven, people don't feel well cared for, and many Americans have no health insurance.

The challenge of transforming health care, continued

There are no easy answers to these challenges. With basically the same health-care workforce we have now, and the same resources, we must find orders-of-magnitude increases in effectiveness and efficiency in order to meet the needs of our patients and society. The challenge is ours: To take what we have and do things very differently.

How health care is seen

It's interesting to look at health care because what it looks like depends, to some extent, on where you stand. From the inside, from the practitioner's perspective, we see ourselves working hard every day to give patients the best care we can. We care, but we're busy. Our patients usually do well, and every year they do a little bit better. If a patient makes it through the door of our hospitals or clinics, we care for them; we don't turn people away. But it's getting harder to do a good job. Staffing is tighter; resources are being trimmed; patients are getting sicker and older; reimbursement is falling steadily. That's how it feels from the inside.

From society's perspective, though, health care looks very different. Costs are way too high and are going up at intolerable rates. Mistakes are made—huge mistakes when seen in aggregate. Also, patients say, "Nobody listens to me anymore." To make matters even worse, many Americans have no health insurance; estimates are that between forty-two and forty-three million Americans are uninsured out of a population of about 275 million. Those are staggering numbers of people, and the evidence is very compelling that without such coverage health status does not reach its optimum. So we have this extraordinarily difficult situation in that healthcare workers are doing the best they can, under increasingly difficult conditions, and yet important needs of patients and society are not being met. It is not an easy time.

Problems as the unintended consequence of success, not failure

We often approach problems by looking for someone to blame; but I am increasingly convinced that this approach is very unproductive in health care. One of the most surprising conclusions I came to after studying these problems for several years is that, in most cases, the problems we face in health care are the result of success, not failure. Most of our present problems are the unintended consequence of well-motivated, successful efforts to solve previous problems. Let's take a look at our concerns of cost, quality, and loss of a patient center, one at a time.

The link between cost and capability

First, there is the concern for cost. We are nearing resource limits in health care. What is the cause of the cost problem? At first it seems easy to assign blame; to assume the cause must be waste, or fraud, or abuse of the system. But careful studies have shown the reason for the problem of high cost lies in another direction. Our high costs today are directly related to a major problem of an earlier era in health care: the problem of limited medical capability.

In years past we couldn't do anywhere near as much as we can today. In 1957, when my father graduated from medical school, much of the care we are able to offer today as a matter of routine did not exist. For example, in my specialty of cardiac surgery, the first operation using the heart-lung machine occurred in the year

The link between cost and capability, continued

of my birth, 1953. The first heart valve was implanted in 1960. The first coronary artery bypass operation was performed in 1965, the year that Medicare was created. It wasn't until the 1960s, and even the early 1970s, that hospitals began developing intensive-care units, coronary-care units, and other specialized units that we take for granted today. Administering many of the drugs that we have today also began then. In short, since World War II there's been a phenomenal expansion of medical capability.

Health care has become more complex—we have added exponentially to what we can do—but costs have increased proportionately as well. Our very success at solving an old problem, limited medical capability, has unintentionally created a new problem, high cost. We have overcome limits of one kind but have reached limits of another kind—limits of the societal resources that can be dedicated to health care. This is not to say that we should ignore waste, or fraud, or abuse. But a much greater challenge lies at the center of the cost problem. And the more successful we become at solving capability problems—the artificial heart comes immediately to mind—the greater our cost challenges will be.

There are only two general solutions to the cost problem when viewed from this perspective: either we do less for patients, or we find new ways to do more. For me, there is only one acceptable answer.

Viewing the issue of quality

Next let's consider the issue of quality. There is an understandable concern for quality today in health care. The recent Institute of Medicine reports on medical error are startling. How can we, with a clear conscience, tolerate 100,000 people a year dying in this country due to medical mistakes? The answer is we can't! Recent studies have questioned the magnitude of the IOM conclusions, but even one preventable death is too many. How shall we think about his problem? Just as in the case of cost, I believe that our quality concerns today are grounded in success, not failure. Let me explain.

Concern for quality is not new in health care. What is new is an emerging understanding of the importance of system-based approaches to quality, approaches that have proven themselves in other industries that are now achieving quality at six-sigma levels.

The time-honored solution to the concern for quality in health care has been a professional commitment to *individual* responsibility. When I was a medical student in the 1970s I was told: "Don't trust that laboratory report; look through the microscope yourself. Go do this, then go do that, because there is no greater responsibility than caring for the life of another person." Well, we took that to heart, and have created the best system that can be made by self-reliant, individual caregivers doing their utmost not to make mistakes.

We have succeeded in developing deep cultures of individual responsibility that were appropriate in an earlier era, but paradoxically are now unintended barriers to the adoption of system-based approaches necessary for taking quality to even higher levels. We have no familiarity with such approaches and resist them. We have been trained to be distrustful of anyone except ourselves. Our old medical philosophy said, "This is the most important task you could ever have. Do not make a mistake."

Viewing the issue of quality, continued

A culture of mistrust, shame, and blame is the other side of this intense culture of responsibility. In other industries, such as aviation, which I'll cover more directly in a little while, the approach was and is very different. It is assumed that people *will* make mistakes, and systems are built to keep what is understood as inevitable human error from causing adverse events. So, again, our challenge is not based in failure or lack of concern about quality, but rather it is an unintended consequence of success from an older perspective.

Loss of a patient center in our work

In 1924 my grandfather began his medical practice. When a patient came to him he knew that patient by name. He knew the patient's family situation. He knew that he might not get paid but that that would be all right. He knew that maybe the patient was coming in because he and his wife were fighting, or were worried about their finances, and not necessarily because of stomach pain. My grandfather also knew that he could help craft a solution and that when the patient left he would feel truly cared for in a holistic way. The term "patient-centered care" describes this type of relationship.

We have a wish, in health care today, to regain this kind of relationship, this sense that health care is really in the service of the *person*. We hope that health care will be in the service of repairing *a disrupted life*, not just in fixing a metabolic abnormality or correcting this or that condition. The wish is for health care to truly be in the service to the *whole patient*. That's what the term "patient-centered care" means. But care like this is hard work, and it gets harder every day. Our systems no longer support it.

Why have we lost our focus on the patient as a person? Again, I think this is yet another example of an unintended consequence of successfully addressing a previous problem. This problem of the loss of a "patient center" in our work became prominent as care itself became more and more complex. The challenge then was mastering this increasing complexity and sophistication of medical science. The solution was specialization. The unintended consequence of this solution was a gradual emphasis on treating medical conditions, not people. Health care became specialized, and now we have a heart doctor, a lung doctor, a brain doctor, a cancer doctor, but we never seem to have the *patient's* doctor anymore. Our success in solving the problem of complexity created a new problem. Now the patient says, "No one listens to me anymore." We are learning that great care of organs and diseases sometimes feels terrible to the whole person. Our past success again set the stage for our present discomfort.

A legacy from the 1920s

The same paradox is true of the way health care is organized, and of our usual work patterns. Most of the organizational structures of our healthcare system are legacies from the 1920s. They are the result of well-intentioned efforts by healthcare professionals who were attempting to organize care in the best way they knew at the time. Those ways of organizing care were, for their time, a great advance. But, here we are today, with those same organizational structures and those same work patterns. In many cases those organizational structures and work patterns no longer help us. In fact, they often stand in our way. But because that is how we were

A legacy from the 1920s,
continued

trained, and for the most part how we still work, it is hard for us to even imagine doing things differently. It is not easy to step outside of the cognitive models that we use to define and organize our world. But we must. We can do better.

Commitment and care as vital forces

I'd like to shift, for a moment, to some positive things in health care. There are some very valuable resources in health care that are often forgotten and left unmentioned. The most important is the reservoir of care and compassion that *does* exist in the healthcare workforce. I marvel as I watch the nurses in the Intensive Care Unit caring for our patients, for the families involved, worrying over their needs. The same is true for every healthcare professional. It's not fashionable to recognize it, or even acknowledge it right now, but the care and concern are there. This passion is crucial. Its power should never be underestimated. Throughout the history of health care this commitment to give patients better care has been the most important driving force for improvement, far more significant than economics, regulations, or any other incentive. I see this commitment as a vitalizing energy force capable of transforming our healthcare system. Our task is understanding how to harness it productively.

The notion of customer

I'd like to say something about the idea of customer service and the term "customer." Now, I'm often a customer, and I am also a healthcare practitioner. I would agree with many who say that in health care there is a lot to be gained from thinking about ideas of customer service in business. I've never been a business owner, and I don't understand the notion of customer exactly from that perspective. But as best as I do understand it, "customer" is an important word in business, and *fundamental* to the concept of customer in business is the idea of respect for the individual.

But I also want to say, as a healthcare person, that there isn't a perfect fit between the term "customer" and my profession. We often feel a sense of responsibility and connection to our patients that transcends what I understand the customer/business relationship to be. It's a deeper relationship and one that is much more personal. In a sense it is similar to the responsibility that one feels toward a loved one. So I have a little difficulty with the word "customer." My relationship with my patient means even more to me than that. But for many people the word helps to define relationships of service, so I don't mind it too much. We just need to remember how special, sacred really, this relationship must always be. The great surgeon and educator Francis Moore said it succinctly: "The fundamental act of medicine is assumption of responsibility." This human connection, of responsibility for another, is at the center of health care.

Circle of identity

Related to this sense of responsibility for a patient is a concept that one of my friends, Martin McKneally, calls the "circle of identity." His thinking has helped me understand what my role is when I ask for consent for a procedure or recommend a particular course of treatment. By "circle of identity," Martin means the way in which we grow to become fully responsible, well-functioning individuals.

Circle of identity,
continued

To his way of thinking, when we are born our "circle" is rudimentary, kind of empty, and by the time we get to around twenty-one or twenty-five that "circle" is pretty well fully formed, although fluctuations do occur. Over the years those who teach us, our parents and others, step in and out of that circle, helping us when we need that help; but over time they're always stepping back, with the goal of making that circle intact, complete, bright and vital.

In normal business relationships, customers are presumed to come with a fully intact functioning circle of identity. The world of business presents people with choices; it expects them to make rational, informed decisions, and live with the consequences.

In health care we have made mistakes in both directions regarding the circle of identity. In earlier eras health care grievously stepped too far into the patient's circle of identity, depriving the patient of the dignity of making independent decisions. But today we are at risk of making mistakes in the other direction, of assuming that a person can be a rational decision maker when that is just not possible for them. Very often the thing that brings the patient to us also works on the circle of identity. For example, if a patient suddenly has a heart attack, that patient may not be able to decide clearly and rationally about risks, options, and whatever else. Part of the patient's circle of identity may become incomplete; although the circle may still very much be intact in its majority, there may, nonetheless, be gaps.

What my friend has helped me understand is that it is my professional duty to step forward just enough to make that circle whole, to lend a little of my own identity to that patient for a while. I can do this in different ways. I might say: "Here is what I think the facts are." "Here is what I would do for myself or a member of my family." "Here is what I recommend to you." I might also ask, "What other information can I provide for you to help you make a good decision?" The reality is that I'm steering a lot of that interaction, and if I say otherwise I don't think that I am being totally truthful. Our present laws and procedures assume otherwise, but the reality is more complex. The reality is that the patient's circle of identity often is incomplete; and health care, almost uniquely in our society, requires practitioners to step forward and help repair that. It is our ethical responsibility to step forward. But equally we have the responsibility of always working ourselves out of that circle, of making that patient whole again.

The social contract of health care

I think this vulnerability, and our professional responsibility because of it, is what underlies what has been called the social contract of health care. Sam Thier has written about this eloquently. The social contract of health care says, in essence: *Society grants to the health professions the privilege of caring for the sick, and many other honors and rewards, in exchange for the promise that the welfare of the sick will always be held before all other concerns.* These are the absolute terms of our relationship with those we care for. We violate this social contract at our great peril, as managed care is learning the hard way. Any other relationship does harm to the essential soul of our profession and will not be tolerated by society. This basic principle must lie at the heart of any new system of care.

Importance of self-care

I have been discussing some basic principles that underlie what we do in health care: ideas of reciprocity, patient service, the circle of identity, the social contract of health care, and so forth. I also think that we need to add self-care and peer-care as essential principles in a healthcare system.

We all need to take on the responsibility of caring for ourselves; it is an essential first step. Dr. Herbert Benson, as you may recall from his article in the previous issue of this *Journal*, thinks of a healthcare system as a three-legged stool, with procedures and pharmaceuticals being two of the legs, and self-care being the third leg. It is worth thinking about self-care, because we often shortchange this leg.

The concept of self-care is deceptively complex. Self-care is not the norm in our present systems. Instead, present norms tend to keep "ownership" of care within the system itself, in effect keeping patients at a distance from the information and resources they need to care for themselves. Again, I don't believe this has been deliberate. It's just the way it has worked out. But the result is that we have unintentionally disenfranchised the most valuable workforce in health care, our patients themselves. Not only do people want and need the information necessary to manage their own illnesses, there is a practical side to it as well: we will never be able to care for everyone as well as we would like unless we all step forward as true partners in that care.

There is another dimension to self-care that begins very close to home: how healthcare practitioners work and live. The news here is not good. As a doctor, I exist in a culture of tremendous overwork. Doctors, especially in the training years, may stay up all night and still go to the operating room the next day. We may start our days before dawn and get home at 8:00 or 9:00 at night. Nurses and other caregivers are being asked to do more and more with less and less. This tendency toward overwork is epidemic in American society as a whole and is especially true in health care.

A colleague of mine once asked me an interesting question that really struck home for me: "How can we take good care of our patients if we don't take better care of ourselves and each other?" Her point really is worth considering. We are trained and acculturated to look outward, toward our patients and our work, and not to look inward at ourselves. We often model in our own life choices that are exactly the opposite of what we hope to accomplish for our patients.

What is quality?

Obviously, when you're working with a patient, or on a person's heart, you want to do the highest-quality work you can. But what does that mean? Just take the simple question "What *is* quality?" It's worth thinking about that for a minute.

I don't know that there's a right answer, but one that I love came from an elderly lady in the second row at a talk that I once gave. The members of the group I was addressing were all senior citizens. Many were eighty or ninety years old. I asked the group to think of things in their life that exemplified "quality." The gentlemen talked about certain kinds of things they had made, and the ladies spoke of certain things they had baked, or sewn, or of relationships. At first they were very traditional in their gender divisions. They agreed that quality was hard to define, that you just

What is quality? continued

sort of knew it when you saw it.

But then one lady put her hand up. She said, "I think it's just paying attention to what you're doing." I was really struck by that. I said, "Tell me more. What do you mean by that?" and she said, "Well, you just look carefully at what you're doing, you reflect about it, and then you try to do it better." The group agreed. I thought, "Well, that's pretty interesting."

We eventually decided that *quality is less a property and more a way of being. It is what happens when you pay attention to what you're doing, reflect on it, and try to do better as you go along.* I think it would be hard to come up with a better definition.

Linking quality with learning into an "improvement loop"

I have thought about her answer a lot. What I eventually understood she meant is that there is a relationship between quality and learning. I think she was saying that quality constitutes part of a learning journey. I eventually linked that thought up with another idea about learning that was expressed by Gregory Bateson. Gregory Bateson was married to Margaret Mead, the famous anthropologist. He was a highly original thinker, back in the 1950s and 1960s, one of the early theorists in the field of cybernetics. He said, "All learning depends on the ability to detect difference." His point was that if everything seems the same, if you can't detect any difference between one state of being and any other, how could you possibly learn?

So, if you put together Bateson's interesting quote that learning depends on the ability to detect difference, and my audience member's idea that quality is a learning journey, what you get is something I like to think of as the "improvement loop." Here is how it works. You have some process, and you want to build quality into it. So you add learning. You measure something about the outcome of your process; you compare it for a difference with previous outcomes; you reflect on the difference; and then you try to do better. You keep running this over and over again, and pretty soon what you get is "quality." Easy as pie.

What is our level of self-awareness?

Clearly, though, you need some level of self-awareness to even know you have a process going. In health care, many people aren't at this level of self-awareness when it comes to systems and processes. They're just doing something. If you stop and ask them how it's happening, they'll just say, "Well, I'm just being the best doctor or nurse I know how to be." They don't really understand what they are linked into and that everything that they do is rising or falling depending on the environment that they are in. So, as a first step, you have to have some degree of self-awareness.

Measurement and quality

As a second step, you also have to have some sort of measurement in place. Most healthcare systems do not have outcomes-measurement systems, except in very crude forms. The systems that do exist are more often used for judgment and monitoring rather that as learning tools.

Within the vast body of what we do there is a relatively sizable domain known as "evidence based" practice: treatments for which there are reliable, scientific evidence

Measurement and quality, continued

that if you do this, that will happen, and so forth. It is important to remember that this body of knowledge changes over time. When I was a medical student one of my teachers said, "Half of what I'm teaching you is wrong. The problem is, I just don't know which half." It's evolutionary, I guess. Still, at any given time, there is knowledge that should be put to use. It is interesting that, of all that we do in medical practice, most estimates are that only about 15% to 20% has a true rigorous evidence base, a scientific basis. A good deal of the rest of it is sort of experiential.

There's been quite an interesting push over the past ten years or so in health care for more evidence based practice. Let me give you a very concrete example. There's an overwhelming body of evidence that if you have a heart attack you need to be on a class of drug called a beta-blocker. Patients on those drugs have a dramatically lower risk of having another event or of dying. There's clear evidence for it. It's very well known. And yet when studies are done to see, at any given hospital, what percentage of heart attack patients went home on a beta-blocker, it's a remarkably lower number than you would wish it to be. Somewhere between 60%, 70%, or maybe 80%, but nowhere near the 98% or 99% that it ought to be. A huge part of this is because, until recently, there's been very little measurement of how many people in hospitals actually went home on beta-blockers after heart attacks. It was just assumed that good doctors would make sure that their patients were on beta-blockers, and to the extent that good doctors could make sure good things happen, patients were on beta-blockers. *We need to measure how procedures that are known to work are being applied.* This process type of measurement is very often not being utilized in health care.

So we need better systems of measurement. But we need even more than this. We also need to have some sort of structure that affords an opportunity for reflection at a system level. In health care, with lots of different people involved, there's no way we're going to be able to talk or think effectively about how to change what we are doing unless we get them all together somehow. And so thinking about measurement takes us back again to the idea of optimizing the social architecture of health care.

Self-conscious and unselfconscious forms

Let's touch upon some philosophical origins and roots. Christopher Alexander, an architectural theorist at Berkeley, authored three books: *Notes on the Synthesis of Form, A Timeless Way of Building,* and *A Pattern Language.* Alexander, interestingly, is not widely known in the architectural world, but in the world of computer programming, especially object-oriented programming, he's considered something of a god. Alexander has an interesting idea that relates directly to measuring and improving what we do in health care by optimizing our organizational structures. He says that human beings create structures or "forms" in at least two ways, which he calls "self-conscious" and "unselfconscious" form making. In "unselfconscious form making" we engage in the kind of direct creative work involved with making, say, a primitive mud hut. With the other type of form making, which he calls "self-conscious form making," humans participate more indirectly, for example by contracting out to someone like I. M. Pei to build them something really fancy.

Self-conscious and unselfconscious forms, continued

Alexander says we were always learning and always improving when we were involved with unselfconscious form making, because the result was immediate. You'd build a hut, and then you'd go live in it. If it rained, and the roof leaked, you'd daub a little mud up there, and if that didn't work, you'd build your next hut in a way that remedied the leak. However, with self-conscious form making, less of a connection exists between the architect who designs the building, the builder who raises the structure, and the people who eventually live in it.

Alexander suggests that the result of engaging with "forms" in an explicitly self-conscious way is often this: we make terrible buildings. Alexander says, "In unselfconscious form making, you do it, you live with it, you change it. This happens almost subconsciously. But it's different in our modern world. Where things get more complex, the connections don't happen."

So am I suggesting that we simply go back to simpler times? Of course not. The reality is that complexity is here to stay. We just need to be smarter about dealing with it. Alexander is not actually criticizing complexity. He is pointing out that increasing complexity has separated us from an immediate and intuitive knowledge of the cause-and-effect relationships of our work. The way we deal with complexity in organizations has effectively decoupled our natural improvement loops. To build quality into our work we need to get those loops working again.

Using tools to make differences visible

I've talked about how improvement loops bring together doing and learning. And about how learning begins with the ability to detect differences. Yet, there are limits to our ability to measure differences. If the result of a process takes a long time, or if changes are subtle, there may be differences that we are unable to see with clarity. So we need somehow to magnify them. That's what run charts, statistics, maps of variance, and so forth were made for. They are tools that magnify small changes. We need these tools. They enable us to view subtle changes clearly and respond to them appropriately. In health care the desire for quality is evident. But the tools needed for seeing differences aren't always well used. We are going to have to begin to use our tools. They are crucial for measuring our processes of care and, in turn, for improving them.

Product and process innovation

When we talk about transforming health care, we are basically talking about innovation, which means doing something that's never been done before. Economists often divide innovation into two types: "product innovation," which is improving the goods or services of an industry, and "process innovation," which is improving the production resources of an industry.

In health care we have had remarkable product innovation, and almost no process innovation. We have better and better drugs. We have better and better operations. We have better ways of monitoring, testing, checking this, and checking that. Yet when we look at *how* the production resources in health care are assembled to create those innovative products, we have really had very little progress for most of the past century.

Product and process innovation, continued

It's not the same in other industries. For example, if you go into a computer plant today and compare it with thirty years ago, you will find that the computers are much better *and* that the production process today bears no resemblance at all to how computers were made thirty years ago. *Both* the product and the process have changed!

In health care, for all the advances that have been made, the "production process" works almost exactly the way it evolved a century ago. It is more complex, certainly, but the fundamental ideas of organization, patterns of interaction, modes of communication, and information flow are structurally just the same. Almost all the knowledge that has been used to dramatically change production processes in other industries has just completely bypassed health care. It's interesting to ask why.

Why has process innovation not occurred in health care?

In most industries, the primary driver for process innovation is cost reduction. Until relatively recently, cost-consciousness has not been part of medical decision making. For most of the past century, for good reasons, medical decisions and cost decisions were kept distinctly separate as a matter of ethical principle. When the Medicare program was created, it was based on cost-plus reimbursement. In theory, hospitals and doctors did whatever was right for the patient, and the system paid for the cost of doing it. When I was a medical student I was taught, "You're not even to think about cost." I remember when the CAT scan first came out and we were trying to figure out when we should use it. We were basically told, "Young doctor, your job is to take care of this patient the best way that you know how. It's someone else's job to figure out how it is going to get paid for." There were no incentives built into the system to keep costs down; in fact, they were deliberately built out of the system, because the intent was to be sure that people got the best care possible.

As a nation we have just finished a ten-year experiment with managed care, which was basically an attempt to rapidly bring process innovation into health care by intense cost-consciousness. I believe the experiment has failed. We hear a lot about adopting ideas from business and industry, and there is a lot that business has to offer health care. But I think we've found out that differences exist between health care and business, and that these differences can be profound. People find it very objectionable to have cost concerns layered over clinical decision making. For fifty years we have enjoyed the results of explosive product innovation in health care, yet because we're fifty years behind in process innovation everything is staggeringly expensive. Yet, when we try to drive cost-consciousness into health care by competitive market forces, we hate what we produce.

So this is our challenge in health care today: How can we drive process innovation in health care, yet not distort the essence, the fundamental spirit, of what health care is? I think the answer is in reformulating the question. The question we need to ask is "How can we help healthcare practitioners actualize their commitment to better patient care by incorporating process innovation as an ethical dimension of their work?" It think the answer to this better question lies in the clinical environments we build around them.

Creating an optimum environment for system improvement

I believe the key is creating optimal environments for an inside-out transformation for health care. This is a far better strategy than focusing on economic incentives, oversight, or regulation. The latter is the tack that we've taken over the last few years, and it's proving to be a recipe for making a healthcare system that patients neither want nor will tolerate.

I think we can make a system that works better than one founded on external pressure, where somebody from the outside is always saying, "You must do a better job." I think that we can, instead, set a stage that allows for the emergence of this natural tendency of people who are working inside the healthcare system. We must create a system that allows these deep feelings of human responsibility, of compassion and care for others, to naturally arise and become active in shaping our patterns of care and interaction. Right now we don't have structures that allow this commitment to translate into reduced error, better outcomes, and so forth. I think that the care and compassion of our healthcare workforce is the best hope for driving healthcare transformation. My belief is that health care will be transformed by the shared effort of people inspired by the vision of giving better care. *What is necessary is a different infrastructure that actively supports the natural inclination of caregivers to do the best job for their patients.*

Productive environments for innovation

Much is known about environments that produce innovation. Innovation has been studied extensively, and there's a huge amount of literature about this. Perhaps the best example of an environment designed specifically to produce innovation is the Lockheed Skunk Works. The Skunk Works developed almost every new fighter jet that our country produced from World War II until almost the end of the century.

Basically, the leadership at Lockheed took a big manufacturing plant and made a little subdivision within it. They essentially said to the employees, "You people are not quite beholden to us in the same way that everybody else is. We're still going to hold you accountable for your work, but we're going to give you some protected time, we're going to give you some undesignated resources, and we're going to let you make decisions right on the spot." And then they took a most important step. The leadership brought the pilots, mechanics, and engineers together and put them all in a single room; they arranged their desks so that they literally bumped shoulders and elbows with each other, creating opportunities for them to interact with each other on a frequent basis. It was in those informal interactions that the different groups all saw one another's ideas on the subject: Pilots looked at the airplane one way, engineers viewed it in another way, and mechanics perceived it in still another way, even though they were all looking at the same airplane. As the different groups talked and interacted, new ideas began to emerge—spontaneously. *That's* where innovation happens: at the *intersections* between disciplines, *between* people who each see what is apparently the same thing from slightly different viewpoints.

What's been discovered is that *innovation occurs in environments where there's a very flat hierarchy.* A rigid hierarchical system tends to discourage innovation. The

Productive environments for innovation, continued

latter is more for running the show once you've got something figured out, but definitely not for figuring out something in a new way. It is known that in addition to this flat hierarchical structure, when aspiring to generate innovation you need a number of elements:

1. You need *multiple viewpoints.* This should be facilitated by specific ways of improving communication, even if it's just bumping elbows with the person next to you.

2. You need *authority to make decisions.* If you have to go through five or six layers of approval to try something new, then nothing is going to happen easily. It is much better to be able to say, "Let's go do this today."

3. You need some *flexible resources.* Interestingly, though, innovation generally happens best when resources are not lavish. The garage in which Hewlett and Packard created their company comes to mind. They were just two people trying to figure out how to do something with what they had, and just a little more. If your resources are too much and too lush, then innovation may kind of retreat or recede from you.

4. You need *clear goals but open means.* Over and over again, innovators at Lockheed had clear and measurable goals. They would say, "We will make a new jet that will fly at three times the speed of sound in two years with this many dollars." It was that specific. But they were careful not to mandate the "how." The "how" is where innovation lives.

5. You need *intrinsic motivation.* It is always the case in innovation studies that people who are passionate about the thing with which they're engaged do a dramatically better job than people who are bribed, beaten, or encouraged in any other way. Intrinsic motivation trumps, by astronomical amounts, extrinsic motivation.

6. You also need *leadership* of a unique kind. Kelly Johnson was the leader of the Skunk Works, and he did two things. First of all, he gave his people a lot of room to do things on their own. He inspired them, gave them support, and helped them believe in themselves. So he led them inwardly, but then he also protected them from the outside. He said to others, "You stand back. These are my people. If you have a problem, you come to me." This type of inward-outward leadership is enormously important and enabled the type of thing that was happening there.

So how does this all translate into health care? Unfortunately, not very well. In almost every category, the attributes known to produce innovation are rare in healthcare organizations. In fact, patterns of organization in health care are usually just the opposite.

Health care: information transfer in a relational context

Now, let's take all this discussion and begin to focus it. Where we are headed is to the bedside of our cardiac patients in Concord. I am going to tell you how we are taking all of these ideas and giving them form in our program there. But to understand what we are trying to do in Concord, there are two more ideas that we need to consider. These are about the central role of information transfer in health care and about the small-scale organizational structures, which Paul Batalden calls "micro-

Health care: information transfer in a relational context, continued

systems," where the real work of health care actually takes place.

Health care is, in some sense, a process of information transfer in a relational context. This may sound kind of strange, but look at what I do. Yes, I operate on people. I talk to them. I examine them. But I'm also involved with something else. I'm taking in information from patients about how they are doing, and I'm pulling that in with other information I'm learning about their condition from tests, and the like, and I'm adding to that base what I have from my medical training. I may supplement this by looking things up, or researching things in other ways, and then I apply all this information, with other resources, to try to make the patients better. This is all done in a context of relationships; it's information transfer in a relational context.

To me, the efficiencies we are seeking in health care will come from making the relational contexts function as "information superconductors." This implies that virtually no energy at all is being used to make the information flow effortlessly throughout the entire system. It also implies that the transfer of information is absolutely precise and complete, and is communicated to every single part of the system involved. I think that if we direct our efforts toward the efficiency and precision of information transfer, all the other things, such as cost and quality, will follow.

Seeing the microsystems

Paul Batalden and other innovative thinkers at Dartmouth Medical School have been very interested in what they call the microsystems of health care. People in health care have tended to think about systems in terms of health plans, hospitals, and other larger organizational structures. But in reality, according to Dr. Batalden and his colleagues, the *functional unit* of health care turns out to be something else. The functional unit, or "microsystem," is the small group of people and resources that are brought together on a daily basis around the needs of a particular kind of patient. What is interesting is this: If you look at the organizational chart for most healthcare institutions, these microsystems are nowhere to be found. They're almost invisible.

Learning from architectural theory

Christopher Alexander, the architectural theorist I mentioned earlier, looked at various patterns found in nature and then looked at architecture. He said, "You know, there are certain patterns that just make people feel good. For example, if you are looking out through a number of windows at a garden and the sun is shining in, you feel great. On the other hand, if you're in a great big room that has no windows, or maybe one tiny window on the north wall, you feel terrible." Alexander went through various patterns that recur across the world, and across time, and he developed what he called a "pattern language" for architecture. He suggested that you could take these patterns, put them together, and construct buildings out of them. As it turns out, his theories didn't work very well in architecture, but they worked great in computers. The entire field of object oriented programming is built on packets of reusable code that are pieced together in the way Alexander envisioned for his buildings.

Regardless of how the concepts are used, I regard Alexander's thinking about

Learning from architectural theory, continued

patterns to be hugely important. In the book *Notes on the Synthesis of Form*, Alexander puts up a random pattern of dots connected by lines and asks, "OK, how would you cluster this if you were going to group things together?" Alexander suggests that you should group things by taking those that are richly connected in some way and putting a boundary around them. Where there are sparse connections, you recognize that as a place where boundaries should cross.

What does this have to do with microsystems? The idea is that if you have a certain set of people who work together all the time, then you ought to put some sort of organizational circle around them. The boundaries should cluster together people with a common purpose. The individuals forming the circle's interior, then, have to match one another in some kind of significant way.

Organizational circles in health care

In health care almost all of our organizing circles were created back in the 1920s, when scientific medical practice first came into being. We've changed almost everything that we do since then *except* for our circles, except for our way of organizing healthcare relationships and processes. When I look at most healthcare organizations, I see groups of people who come together every day to care for certain kinds of patients; but there are rarely well-defined organizational boundaries around them that match the way care is actually given. The nurses report to nursing leaders, the doctors are over here, the physical therapists are over there, and management is somewhere else. We pull all these people together around a patient, and then when the immediate work is done, everyone goes back to their own departments. The old boundaries don't match the way we work anymore. They don't help us, and sometimes actually get in our way.

Recognizing the circle, the microsystem, at Concord Hospital

At Concord Hospital we are working to build circles around people who do similar things—not similar in their exact professions, as, say, doctors or nurses, but similar in that they are all coming together to care for a certain kind of patient, say, a heart patient. Now, wouldn't it be nice if we could put a boundary around all those people so that, in some metaphoric way, they became similar to the "open room" with the warm sunshine coming through it that Alexander spoke of? We need to begin to pull people together like this.

I think that a fundamental design principle for healthcare systems should be this: *Patterns of organization should reflect patterns of interaction. And patterns of organization should match patterns of patient need.*

The key to healthcare transformation is optimizing the patterns of interaction, and the information transfer that occurs within those patterns, by making them respectful, effortless, and precise. This is mostly about facilitating how people work together. I think that visualizing our relationships, recognizing the rich connections and the sparse connections and getting the boundaries right, is where we should begin. I believe we should consciously seek out the "pattern language" of health care, creating a new "social architecture" that helps us do our work much more effectively and efficiently.

Our work at Concord Hospital

And that leads me, finally, to the work we have been doing during the last year and a half in our program at Concord Hospital. First, let me tell you a little about Concord Hospital. The quality of the hospital and its nursing and physician staff is unusually high, and there is a very strong ethic of wanting to do the right thing for the community. The CEO of the hospital, Mike Green, is very gifted and forward thinking, and so is his leadership team. Mike and others decided several years ago to bring more-sophisticated services to Concord. Patients, at that time, were leaving the community for treatment for complex conditions, such as cardiac care or cancer treatment. He wanted to expand the services that could be offered locally.

He started out by setting a multi-year plan and began by strengthening the cardiac program. A major part of that was the creation of an open-heart surgery unit that was developed through a partnership between Concord Hospital and Dartmouth Medical School. Dartmouth has an extraordinary reputation nationally for quality. The cardiac surgery program came together under the direction of Dr. Steven Plume, who at the time was the president of the Dartmouth-Hitchcock Clinic. Dr. Plume wanted to bring the best of Dartmouth to Concord.

I want to point out that the entire region of northern New England has the best outcomes in cardiac care of any region in the country; the only one that is equal to it is New York state. This was not always the case. About ten years ago, long before the creation of the Concord program, a collaborative quality improvement initiative was begun in northern New England that involved all the programs providing cardiac care in Maine, New Hampshire, Vermont, and northern Massachusetts. That group was called the Northern New England Cardiovascular Disease Study Group.

The doctors, nurses, and administrators in these different programs agreed to meet four times a year to discuss how they were doing in terms of caring for their patients. The teams throughout the area visited one another. They observed the care given in the operating room. They videotaped operations. They exchanged information. They agreed to use a standard database to measure outcomes and to share the information, although the information itself was utilized in a blind fashion. The information was simply being shared with the goal of improving quality across the region. During this time there have been dramatic reductions in deaths and complications from cardiac care. Now a little bit of that reduction has occurred as a national trend, but the region has surpassed that by a significant degree.

With this heritage of Dartmouth and the northern New England region in general, the Concord program was really an outstanding program right from the outset. I mention this because I am going to tell you about how it became even better.

Getting started at Concord Hospital

In the fall of 1999, we initiated a transformation effort in our program. We took a very good program, but one basically operating within a more traditional healthcare framework—with doctors doing their thing, nurses doing theirs, and so forth—and shifted the way it was being run. We initiated instead a much more collaborative care process, and made *collaborative rounds* the centerpiece.

Collaborative rounds

You may not stop to think about it but there are quite a number of professional disciplines that interact around the care of an open-heart surgery patient. In our program about sixteen different disciplines come together around our patients. Doctors are one discipline. Nurses are another. There are respiratory therapists, physical therapists, social workers, occupational therapists, spiritual care people, and on and on.

This extended team is really quite large, and the challenge of having this many people communicating effectively with each other and with patients and families is quite complex. In the traditional mode of operation, caregivers interact with the patient and each other at various times during the course of the day. Team members communicate in many ways: Some may write a note in the patient's chart, some may make a telephone call, or page, or talk to other team members when they happen to see each other over the course of the day. Some healthcare providers might talk to the patient or family and say one thing, and others something else, and so forth. Despite good intentions, it's actually more a variably connected amalgamation of individual efforts rather than real teamwork.

We thought about this, and decided that we wanted to do better. We decided that we could improve how we functioned and communicated if we came together at *one* time instead of at different times across the course of the day. That proved to be very difficult at first but eventually we were able to get people to adjust their schedules so that we could do that, and we began something that we now call our "collaborative rounds process." It has five elements: (1) Interdisciplinary, with everyone present at the same time, (2) Patient and family included as part of the care team, (3) Respectful, open environment with flat hierarchy, (4) Consistent pattern of communications and decision making, and (5) Specific attention to identifying "system glitches."

Advice from others

Once we got people used to meeting together each day, we began to work specifically on the precision and completeness of our information sharing process. We went outside our industry to Jeffrey Brown, an aviation safety expert, and others with expertise in team-based communications and said, "Help us. We have started this collaborative process, but we don't know quite what we are doing. Could you teach us how we can improve the way we exchange information?" We took a number of new ideas, modified them to fit our needs, and added them to what we were doing. With these insights, and with the entire team working together, we developed a communications protocol that is consistent with the best teachings in cognitive psychology and human factors research, yet is applicable directly to our work caring for patients.

Changes in patient care

We made other changes in the program as well, which at first might seem a little bit paradoxical. For example, we started doing less for our patients. We started monitoring patients less with invasive monitoring lines that are so often used after surgery. We started taking lines and tubes out of the patients sooner. We started getting breathing tubes out really early, as soon as the patient comes out of the operating room. We started really minimizing the number of medicines that we gave people.

Changes in patient care,
continued

Broadly speaking, it's really a philosophical change from the way we grew up in heart care. The old idea was that the heart patients were the sickest patients in the hospital. We'd kind of say that the patients were *sick until proven well*. Now we think of them differently. We think of them as *well until proven sick*. We now say, "Fine, you had a little thing done to you, and that's great. But all your lines and tubes are out. Get out of bed. Let's walk around. Let's get you back to what you were doing before." It's a very different mindset.

Involving the team in clinical assessments

We began to stress the value of what we call "integrative clinical assessment" of our patients more than our traditional invasive measurement of how they are doing. I might ask the nurse, "What do you think about this patient?" and the nurse might say, "He looks sick to me," or "He actually looks pretty well, and when I think about it, he looks well because he's warm and he's pink and he's peeing and he's talking to me." We used to measure all this with our lines, and with all sorts of other things. Now we're asking the nurses and the doctors to put more emphasis on their integrative clinical skills, rather than on the measurements that break assessments down into "the pulmonary artery pressure is this, the central venous pressure is that, the cardiac index is this," and so forth. We have tried to step back a little bit and see the bigger picture. Not that the details aren't important. But the context is critically important, too. It is the overall picture that really matters, and that big picture is more than a collection of measurements. We began trying to do all that in this collaborative setting and, absolutely significant, where we have the families and the patients together too. We are trying to build an understanding of how the patient is doing from these multiple perspectives.

Working on the environment for communication

We also tried a number of things to create an environment where people would feel safe and enfranchised enough to give their opinions, even to express concerns or offer suggestions unrelated to their specific area of expertise. I'll give you an example of what I considered to be a true victory of this process. Just a few months ago a patient was complaining of some nausea, and everybody on the care team had a turn at presenting their view of how the patient was doing. Basically, nobody had picked up on the fact that the patient was complaining of nausea, and that the patient was still on some narcotic pain medication, which actually causes nausea at times. Finally, at the very end of the rounds process, someone brought the problem out into the open. Past me, past everyone else, our spiritual caregiver finally said, "Well, I talked to the patient yesterday, and I think her spirits are doing just great. But, by the way, don't you think we ought to stop giving her that Lortab because she's nauseated? Maybe we could put her on some Tylenol instead." This just thrilled me that our chaplain was willing to step forward and make a suggestion for a medication change.

Let me give you another example of how helpful the collaborative process can be. During rounds, a patient's wife mentioned that it was very hard for her to stop smoking. She wanted to stop, so that she wouldn't put her husband at risk. I had six

Working on the environment for communication, continued

or seven people on my care team giving her advice. They were talking with her about their experiences with patches, and with this and with that. It is just amazing when you look at a particular problem from all these different viewpoints. It is like a diamond with all its facets! We're trying to put all the facets together to see a whole picture, to step back far enough from the clinical situation on a regular basis to ensure that we're not lost in all the details, and develop a shared ability to grasp the bigger picture of it all.

The Collaborative Communications Cycle

We call our communications process the "Collaborative Communications Cycle." I would like to describe it to you. First, let me set the stage. We are at the bedside of a patient who, a day or so ago, had open heart surgery. The whole care team has come together. Usually it's about a minimum of eight and a maximum of about twenty people. We just gather around the patient's bed. We have the family, the patient, and all of the healthcare staff present. All are part of the care team. It's interdisciplinary. We view the patient and family as essential members of our team.

Achieving excellent communication is really the center point of the whole process. This requires a consistent pattern of communication and decision making. It's a pattern of information exchange that everybody has now become comfortable with, and it pretty much guarantees that everybody will be involved. We don't miss very much anymore. We work very hard to make this a respectful, open environment, with as much of a flat hierarchy as is reasonable. Although the surgeon is ultimately in charge of the process, we actually step back and let our nurse practitioner be the one who convenes the cycle and moves it along. I envision myself more as a CEO of a wonderful corporation with outstanding executive vice presidents rather than as a sole proprietor (see Figure 1).

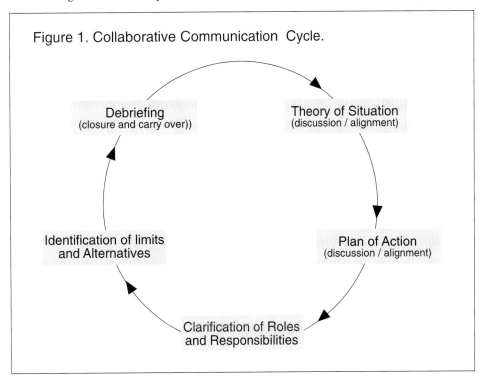

Figure 1. Collaborative Communication Cycle.

The Collaborative Communications Cycle, continued

We start the process by developing what we call "the theory of situation," a term borrowed from aviation safety. At the "theory of situation" stage we sum up just what we believe the situation is, where we think we are with a patient. Somebody proposes a theory of the situation, and then everyone else contributes to confirm or modify that initial suggestion. If we were in an airliner, a crewmember might say, "Well, I think we're twenty miles outside of Boston and that we can start down now," and somebody else might say, "I think we're actually twenty-five miles outside of Boston and there's a big tower that's between us and where we want to be." The captain might respond with, "I'm glad you mentioned that. Let's just double-check where we are."

We might ask somebody like Addie, the social worker, or Judy in spiritual care, or Lynn, from respiratory therapy, "Could you tell us how you think this patient is doing?" And so somebody will offer that, and then we ask, "Does everybody agree with that? Does anybody have any other thoughts about it?" Then we'll take an active period of time to just be quiet and let ideas come out. We even ask questions like, "What are we missing?" or "Does everybody agree?" We'll ask the family questions. We'll especially ask the patient, "Do you agree with that? Do you actually feel that way? Is there anything that you're worried about? Is there anything that we haven't mentioned here?"

When everybody's in agreement we move on to the next step, developing a plan of action. In aviation a crewmember might say, "Given that we're twenty-five miles from Boston and there is a tall tower just ahead, why don't we stay at the present altitude for three more miles and then safely start down?" In medicine we might state it as, "Given that we are doing pretty well but we have these concerns, maybe we want to change that medicine." Again, we open the conversation up for discussion, and we hope that somebody like Judy will say, "Maybe we ought to just stop that medicine." Or maybe a family member will say, for example, "Well, you know, we tried that medicine a year ago, and that didn't work very well." We might say, "Great, let's not use that medicine." And, so, we work with each other, and with the patient and family, to develop a plan of action that is a better plan than what any one of us might have come up with alone.

Next we clarify roles and responsibilities. Often there isn't a great need to say a lot about this because much of our work is role-based, but not infrequently there is some degree of crossover, and clarification is very helpful.

The next step we call "identification of limits and alternatives." We decide together what we expect will happen, what is acceptable and what is not, and what we will do if things don't happen as we would like. There is much less ambiguity. Together we devise alternate plans that can be put in motion if needed.

Finally, we summarize and double-check what we have decided. We end by summarizing: "O.K., we're going to do this, this, this and this. Addie's going to do this and this. We'll reconvene tomorrow. Everybody agree? O.K., great. Mr. Smith, do you agree? Any concerns that you have? O.K., fine." Then we go on to the next patient.

The next day when we come back, we begin where we left off. The person who

The Collaborative Communications Cycle,
continued

summarized the day before will say, "Mr. Smith, good morning. I'm going to review for you what we planned yesterday. Yesterday you still had your chest tubes in. We were going to take those out. It looks like that has been done. How did that go? We were going to get you up and get you to the shower. Did you get your shower? That's great." The talk will go on about a number of concerns. This discussion will form the beginning of a new theory of situation, and then the cycle gets under way, and repeats each day.

Very important, we also interrupt this process and talk about actively catching errors, right on the spot. As problems surface, and of course they do, we use them as a lens through which to look at root causes more deeply within our system. We try to recognize and take note of problems when they occur so we can address them specifically later at a system level. We call these problems "system glitches." The word "glitch" is very effective for us. It is a very easily understood, non-threatening name that allows people to focus on issues without becoming defensive.

System rounds

The other thing that I want to mention is that we also try to get the same group of people together one afternoon a week for what we call "system rounds." We call the collaborative morning efforts the "morning rounds," which are really "patients rounds." But one day a week we sit down and make our collaborative system itself the patient. We probe it. We think about it. We talk about what we are doing and what we could do differently. A lot of our positive work comes out of these weekly conversations.

Results

Early in our work we developed a set of quality indicators that we wanted to measure and track in our program. The dimensions of quality we follow are clinical outcomes, patient/family satisfaction, and quality of work life.

We measure our clinical outcomes by participating in a shared outcomes database with all other open-heart programs in northern New England. This database includes a statistical risk-prediction model that permits us to compare actual clinical outcomes to those predicted for the region for our particular patient population. The results of this comparison are shown in Figure 2 for consecutive patients since the program began. At the arrow, the changes described in this paper were instituted. A significant and sustained reduction in mortality has been observed.

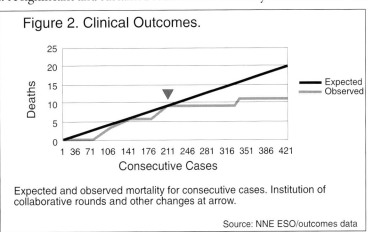

Figure 2. Clinical Outcomes.

Expected and observed mortality for consecutive cases. Institution of collaborative rounds and other changes at arrow.

Source: NNE ESO/outcomes data

Results, continued

We measure patient/family satisfaction using a nationally standardized patient survey (Press, Ganey, and associates). Traditionally, the happiest patients in the Concord Hospital have been mothers having babies. Now, however, the happiest patients in our hospital, happier than mothers having babies, are the open-heart patients. Our patient satisfaction outcomes have also been recognized nationally as outstanding.

In February 2001, we conducted a quality-of-work-life survey of team members, comparing the "collaborative rounds" process with traditional rounds. We found improvements in all eight categories (see Figure 3).

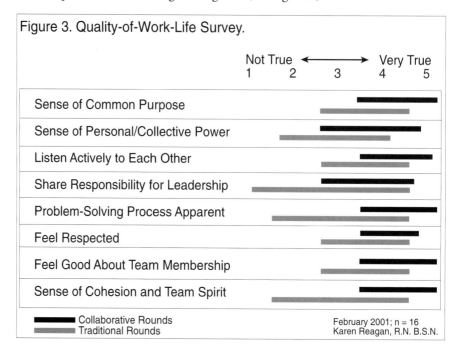

Figure 3. Quality-of-Work-Life Survey.

We are also beginning to receive outside recognition of this work. In the most recent site visit to the Concord Hospital by the Joint Commission on the Accreditation of Healthcare Organizations, the collaborative rounds model received commendation as a national best-practice model.

Conclusion: The concept of relational synergy

It's unlikely that we will ever have better individual doctors, nurses, or administrators, because they are excellent already. Yet, to meet the needs of patients and society, our healthcare system must become remarkably more effective and efficient.

We have, fundamentally, an *individual* healthcare system now. People are immersed in old organizational structures and work patterns that do not facilitate their communication and interaction. The Concord experience shows that something more is possible.

The Concord cardiac program is a special environment for taking patient care to new levels, a living laboratory for innovations in patient care, dedicated to finding new ways for practitioners, patients, and families to work together safely, harmoniously, and effectively. It is a prototype of a new model of collaborative practice that produces "relational synergy." Relational synergy is what happens when parts of a

Conclusion: The concept of relational synergy, continued

system are woven together in new ways, producing resources where there were apparently none before. It is where "the rabbit comes out of the hat."

I'll give you a very brief, final example of relational synergy. In my former community of Wichita, Kansas, a group of practitioners decided that we could do a better job of caring for the uninsured in our community. We learned of Project Access, a program being used in Asheville, North Carolina, and we brought it back to our community and implemented it there. In every community in this country, right now, there are doctors who are caring for uninsured people. There are hospitals giving away care to uninsured people. There are all these parts of the picture then, and every part is trying hard, but none of it is woven together effectively.

Project Access brought together all the parts of the healthcare system. The project linked all the functioning subgroups together so that if a patient needed doctor care, or hospital care, or pharmacy care, or whatever else, he or she could get all of that together. The effect has been that over ten million dollars worth of care in the last year and a half has been given to the uninsured in that community, even though healthcare professionals are doing no more than they already were doing.

Just a small amount of organization, superimposed on work already being done, made something out of apparently nothing. Just as in the Concord experience, Project Access took a pattern of isolated parts and related them functionally, and through this made a system that works in a dramatically better fashion. I think that is what we have to be thinking about and looking for in health care.

Author information

Paul N. Uhlig, M.D., is medical director of the cardiac surgery program at Concord Hospital, Concord, New Hampshire, and associate professor of surgery in the section of cardiothoracic surgery at Dartmouth Medical School, Hanover, New Hampshire.

Dr. Uhlig received his M.D. degree from the University of Kansas School of Medicine, where he received the Thomas G. Orr award as outstanding student in surgery. He completed his residency training in general surgery and general thoracic surgery at the Massachusetts General Hospital and in cardiothoracic surgery at Indiana University. He was also a research fellow in cardiovascular physiology at the Cardiovascular Research Institute at the University of California, San Francisco. Dr. Uhlig maintains an academic affiliation with the University of Kansas School of Medicine–Wichita as adjunct associate clinical professor of preventive medicine.

For the academic year 1996–1997 Dr. Uhlig was the Thoracic Surgery Foundation Alley Sheridan Scholar-in-Residence at Harvard University's John F. Kennedy School of Government, where he studied U.S. healthcare policy. Dr. Uhlig received the degree of Master in Public Administration from Harvard in June 1997. He is presently co-chair of the national health policy committee of the Society of Thoracic Surgeons, and was the lead author of the Society's recommendations for Medicare reform presented before the National Bipartisan Committee on the Future of Medicare.

Dr. Uhlig's professional interests concern transformational change in health care, information transfer in healthcare environments, and collaborative leadership. Dr. Uhlig's present academic work concerns the establishment and study of collaborative environments for innovation and transformational change in healthcare institutions. Prior to joining the Dartmouth faculty, Dr. Uhlig practiced cardiothoracic surgery in Wichita, Kansas, and was the founding president of the Central Plains Regional Health Care Foundation. In February 2000 Dr. Uhlig and Patrick Hanrahan of Wichita, Kansas, received the Mary M. Gates award of the United Way of America for their work with Project Access, a community-based program of care for the uninsured.

Editorial assistance for this article was provided by Erik L. Smith and Laurence Smith.

Six Sigma in Health Care
A Road Less Traveled

Author

Joseph Calvaruso, President and Chief Executive Officer, Mount Carmel Health Systems, Columbus, Ohio

> I shall be telling this with a sigh
> Somewhere ages and ages hence:
> Two roads diverged in a wood, and I—
> I took the one less traveled by,
> And that has made all the difference.
>
> —Robert Frost, from *The Road Not Taken.*

Introduction

At the graduation ceremony for our first group of Six Sigma Guides, which you may call Black Belts, I read the entire poem, *The Road Not Taken*, by Robert Frost. I read it because I believe that a health care organization embarking on a Six Sigma journey is taking a road less traveled by, and we at Mount Carmel Health Systems have decided to take that road. We think Six Sigma already has made, and will continue to make a significant qualitative *and* quantitative difference in health care in our organization. I'm going to share with you our two-year infusion of Six Sigma into the life of Mount Carmel Health Systems.

Mount Carmel is a diverse healthcare system based in Columbus, Ohio. It has three acute care hospitals, an HMO, non-acute care services, such as home health, and a college of nursing. We're a Catholic, religiously sponsored organization, and a member of Trinity Health, Novi, Michigan.

The crisis in health care

We all know there is a crisis in health care. It is multifaceted. There is a shortage of many healthcare workers; not just nurses, but also pharmacists, radiology technicians, surgical technicians, and so forth. As the population ages, we're going to need more healthcare workers, but unfortunately we're getting fewer. At the very least we need to improve things like throughput—reduce rework, duplication, and inefficiencies—so we can make better use of and have a higher productivity from the people that we now have, because we don't have enough.

But there's also another crisis in health care. One out of every five healthcare workers leave their organization every year. This doesn't speak very well of our industry or our leadership. That tells us that something's wrong with what we're doing in health care. A lot of it is because people are feeling so stressed. There's so much rework. Our caregivers are feeling that there's not enough time to deliver quality care, because they're doing so much. As such, we explored Six Sigma as a possible solution.

Anticipating a financial challenge

A few years ago we were also having a challenging time financially. Mount Carmel, fortunately, has never lost money, but the year prior to introducing Six

Anticipating a financial challenge, continued

Sigma we made only $500,000 profit from operations. In that year we encountered four straight months of losses, until we turned it around. And we realized that although we turned around our financial situation the lines were going to cross over into red ink eventually. Our revenues were flat or just slightly rising, but expenses were rising dramatically. We were asking people to do more and more with less. If we continued that pressure again in the next year, it would be more and more and more with less and less and less. Their reward for being able to spin five plates would be that we would give them a sixth plate to spin. So something had to change. We had to make a fundamental change in the way we do business.

A generic, industry-wide problem

But the industry also has a problem. The Institute of Medicine reported that as many as 98,000 deaths a year occur in health systems across the country because of mistakes made in hospitals. It's the fifth leading cause of death, more than from highway accidents, breast cancer, or AIDS, and 7,000 more than deaths from workplace injuries. So we have a fundamental problem. We have processes that are flawed in health care and we need to improve those.

Traditional quality management methods don't take us far enough

We had been using Continuous Quality Improvement (CQI) and Total Quality Management (TQM) techniques and were getting results but we felt that we needed something more. We were at a crossroads. Do we try to accelerate our TQM program or do we try something new? We concluded that what we were doing hadn't prevented the situation we were in. We decided to do something different.

Considering Six Sigma

In early 2000 we started to investigate Six Sigma but weren't able to find a hospital that was doing it. We assumed there must be a few but we didn't find any. Then I read a statement from Jack Welch that Six Sigma was the most important initiative that GE had ever undertaken, that it is part of the genetic code of their future leadership. We thought, "Wow. This is a pretty successful company. This is the CEO of the century. Let's try it. Let's be one of the first in health care to do Six Sigma." Then we thought, "That will take some courage."

Sir Winston Churchill said, "Courage is the virtue that makes so many other virtues possible." And certainly courage seemed necessary in our desire to improve the business and practice of health care. So we gathered our courage and began our Six Sigma journey in July 2000.

Getting started with Six Sigma

Our first step was to find a partner to deliver the needed training and infra-structure that was necessary to begin a Six Sigma initialization. We found a company that met our needs. They understood, had the experience, and had the right sensitivity to the type of leadership that we have at Mount Carmel. So we engaged them and developed a start-up plan. Figure 1, on the next page, is a schematic of our timeline.

The first year was Six Sigma Guide training. We trained twenty Guides in our first wave and twenty-four in our second wave. Our investment was $650,000 and

Getting started with Six Sigma, continued

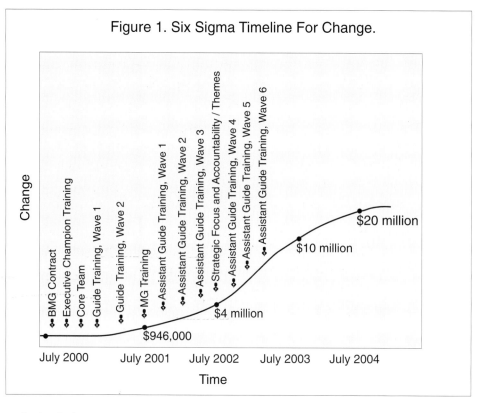

Figure 1. Six Sigma Timeline For Change.

we had a $946,000 return.

In the second year we added Assistant Guide (Green Belt) training and, eleven months into the year, we had $4 million in savings. Some other savings are going to come in and we think we'll end the year around $6 million.

This year (the third) we have a stretch goal of $10 million. We think that is achievable. Our entitlement is probably in the $20 million range. (If you take forty-four Guides, five projects per year, and $100,000 per project, it works out to $22 million. And if you add some of the Assistant Guides, it's probably somewhere around $25 million.) So we think a $10 million goal is certainly achievable.

Improving the focus

Around January or February of 2002 we decided to improve our focus. We wanted to move from a very diverse portfolio of projects to having focused themes. We asked our people, "What are the things that keep you up at night? What do you worry about? What are the major opportunities?" Following are seven themes that they came up with. This is what we're going to start to work on now and move ahead with:

1. Diagnostic throughput. Throughput is probably our most important theme. Operating rooms, for example, don't operate at 100% capacity. The target in health care is 70%. We're running at about 57%. We are turning away surgeries at all three of our campuses because we don't have the capacity, yet we are only running at 57%. Surgical procedures are being performed just a little over half the time that the operating rooms are available. That's not good. We also have a group of orthopedic surgeons who, because they can't get operating

Improving the focus,
continued

room time, are building an orthopedic hospital that's going to compete with us. If we could improve the throughput, not only would we have better patient satisfaction, better physician satisfaction, and greater revenues, we could also stop a potential competitor from doing something because they can't access our operating rooms. The same thing is happening in the CT scan, cardiac catherization, and emergency departments. Mount Carmel has, in one of its hospitals, the busiest emergency department in the state of Ohio. We also have the biggest diversion rate, meaning we turn more squads away from coming to the emergency room. We need to improve our throughput. So we desperately want Six Sigma to help us with that.

2. Bad debt. In February 2002, when we started this, we had $24 million in bad debt, about 2.6% of our revenues. This is not charity care; these are people who can afford to pay and just choose not to. And we have inefficient processes and procedures. So we have developed a very interesting design of experiments to streamline the way we process bills and reduce bad debt.

3. Managed care denials. From July 2001 to February 2002 we had about $5 million in insurance denials. Sometimes it's paperwork we didn't process the right way. These people had good care in our facility. We just didn't get paid because we had flawed processes in the way we sent bills/forms to the insurance companies; a big opportunity for improvement exists here.

4. Improved charge accuracy. Our chief financial officer estimates that 10% to 20% of the services we provided are not charged for and, therefore, we are not getting paid for them. Add all of this up and there's about $100 million worth of entitlement sitting on the table. If Six Sigma helps us to get half or even a quarter of that, it is definitely worth it.

5. Administrative length of stay. Most of our reimbursement is fixed. If we can reduce the length of stay, it helps throughput and improves revenue.

6. Care continuum business projects. We have a home health business, a hospice program, and an ambulance company. There are improvement opportunities here that we are working on, too.

7. Recruit and retain. There are improvement opportunities in recruitment and retention of people, so that we reduce our turnover.

Terminology that fits our purpose and culture

Our logo features the σ^3 symbol. That's for our three S's—Soulful Six Sigma. Caregivers aren't inspired by war-like terminology, so we embrace the soulful (honor every soul with loving service), not the fearful; we infuse rather than deploy; and we have Guides rather than Black Belts.

Quality of work life

There's a Dilbert cartoon that I like. The dialogue says: "This job has taken my dignity, my self-esteem, my creativity, and my precious time on this earth. You've taken all I have; there's nothing left to give." And of course, the boss responds, "The blood drive is next week. This year it's mandatory and a three-pint minimum." Our

Quality of work life,
continued

people in health care feel like that; they feel that they're being bled. They feel that we keep asking for more and more and providing less and less. We have 8,000 employees and 1,200 physicians in our system. We need our people to be inspired. Using Six Sigma for process improvement is one part, but we also need processes that build and maintain a caring and cared-for workforce.

Looking at turnover

We surveyed our people, asking "Why do people stay or not stay at Mount Carmel?" We learned that it is the *quality of relationships* that are the drivers. The number one reason people stay at Mount Carmel is the quality of the relationship they have with their supervisor. The number two reason is the relationship they have with the people they work with. So if we're looking at retention and trying to reduce turnover, relationships are very important.

A refocus of leadership

What makes a great leader? The answers that some of us MBA types may give is: "You have to achieve market share. And you have to have a vision for the future, and articulate that vision." And maybe we'll include all kinds of other "MBA language."

We asked our people, "What do you want in a leader? What characteristics do you want in somebody you report to?" They wanted someone:

- who tells me the truth
- who keeps promises
- with integrity
- who I can trust
- with courage and who's authentic
- who doesn't hide behind their title or their office or their corporate mask.

We all have our corporate persona and our corporate face. Yet people like real people, authentic people, people who show love in the workplace. We all want more love in our lives, and given we spend a lot of time at work, we must get it there. Those are the kinds of people that our people said they want to work for. So we put those two together. This is what people want in a leader and having a great relationship with your leader is the number one reason you stay at Mount Carmel.

We are embarking on a very aggressive leadership development program called *A Journey to Higher Ground*. It is based on work by Lance Secretan, who is devoting his life to reawakening spirit and values in the workplace. One of the major aspects of this program is leadership development. We go away on week-long retreats where we reawaken those kinds of values I just talked about. It's a life-changing experience. We refresh those values that are already in people. We also are creating an environment within Mount Carmel so that when people come back, inspired, the way we're working has been changed so that our people can maintain their inspiration. And that includes Six Sigma, and changing our work processes.

Six Sigma results

I mentioned earlier that turnover in our industry is a little over 20%. In Columbus, it's even a little higher than that because it is a very competitive market. You can get a job in any system. We're all begging for people. And a lot of people jump from one system in Columbus to another just to get recruiting bonuses.

In Columbus, our turnover rate in July 2000 (annualized) was 24%. Because of some of the changes we've made, the next July it was down to 16%. And in June 2002, it was annualized at 13%. In the last four or five months, it's been about 11%. So we more than cut in half the turnover rate in our organization.

It's estimated by people in our industry that a turnover in health care costs about $50,000. This includes the cost of recruiting bonuses and human resource expenses, and the year or two it takes to train people and to get them used to the way the surgeons work at this place and learn their way around the organizations. So cutting turnover in half results in major savings.

We've had an 709% increase in net operating income ($23.1 million) and a 144% increase in net income ($35.1 million). We've had an 8.7% increase in patient revenues ($494 million).

We're very proud of our ability to grow something we call our Community Benefit Ministry. We had a RV outfitted as a doctor's office on wheels. It goes to the people, to thirty places—soup kitchens, shelters, senior citizen centers—where we provide free primary care to anyone who comes in. You often hear in healthcare management today, "If there's no margin, there's no mission." Well, caring for people is our mission. By improving our processes, improving quality, and improving patient satisfaction, we will improve the bottom line and have more funds to do things like this. We increased Community Benefit Ministry funding from $33.6 million in FY 2001 to $41.8 million in FY 2002.

The reason for these improvements is Six Sigma and our leadership development. It includes improving the processes so that people enjoy the work they're doing more, and caregivers can spend more time giving care, touching patients, and being with patients. That's why they went into this line of work to begin with. That's what they want to be and that's what they want to do. Six Sigma is helping us to do this better.

Author information

Joseph Calvaruso is the president and chief executive officer for Mount Carmel Health System (www.mountcarmelhealth.com). Mr. Calvaruso previously served as the president and CEO of the Mount Carmel Health Plan (HMO) and as the senior vice president, Managed Care Services, where he was responsible for managed care activities.

Prior to joining Mount Carmel in 1985 he was a senior consultant for Peat Marwick, St. Louis, MO, and a health planner for Health Systems Agency of Summit, Portage County, Ohio.

About this article

This article was developed from a highly rated presentation at the International Society of Six Sigma Professionals (ISSSP) leadership conference in June, 2002. For more information on the association, you may visit their web site at: www.isssp.com.

Process Design And Management: The Path to Organizational Transformation

Author

Sr. Mary Jean Ryan, FSM, President and CEO, SSM Health Care, St. Louis, Missouri

Preface

SSM Health Care was one of the first health care organizations in the United States to implement continuous quality improvement system-wide in 1990. The system owns, operates, and manages 20 acute care hospitals in four states—Missouri, Illinois, Wisconsin, and Oklahoma, and three nursing homes. SSMHC has nearly 5,000 affiliated physicians and 20,000 employees. SSMHC also owns an interest in two managed care organizations, Premier Insurance Agency in Wisconsin, and Community Care in Oklahoma.

Introduction

In 1999 SSM Health Care received the Missouri Quality Award. We were also the first, and only, health care organization in the country to merit a site visit from the Baldrige National Quality program. If you've ever gone through a quality award application effort, you know that it is a humbling process. First, you must analyze and carefully document your own strengths and shortcomings with regard to quality. Second, and even more challenging, you must welcome outsiders inside, so that they can suggest all of the places where your organization could make improvements.

Any organization that is willing to put itself through the painstaking work of self evaluation and then allow others to scrutinize its facilities, must be committed to quality improvement. When SSM Health Care began the quality journey 10 years ago, we really had very little idea of the extent of the commitment we were taking on. At our annual leadership conference, in May of 1990, we proudly and publicly announced that we were launching a continuous quality improvement (CQI) effort. But, during those hectic early months of implementation, some of us on the system's team of senior executives privately began wondering if instead of having committed ourselves to CQI, we should have just had ourselves committed!

Over time, though, we have embraced the never-ending task of our quality improvement effort and I want to share three things with you. First, I will reflect briefly on the importance of continuous improvement for organizations committed to building quality cultures. Then, I want to look at some lessons that SSM Health Care has learned about process design, process improvement, and process management in the past decade. And finally, I will offer some thoughts about why I believe process improvement leads to the transformation of an organization.

In the beginning, there was some culture shock

When an organization takes up the challenge of shifting its entire culture to one of continuous quality improvement, it enters very unfamiliar territory. For example, within SSM Health Care, in our pre-CQI days, many of us, on both the

In the beginning, there was some culture shock, continued

clinical and management sides, thought that we did things quite well. When we moved into the new arena of quality improvement, however, we began to wonder if we actually did anything nearly as well as we could. We discovered that when we were subjectively judging our performance, we always gave ourselves high marks for our accomplishments. And we always had very good reasons why we didn't make progress in other areas. We also were in the habit of using health care industry standards as a way to compare ourselves to other facilities. And—no surprise—the comparisons were almost always favorable to us! In the new arena, we began to doubt if anyone in health care was thoroughly competent. And, we questioned whether our industry standards measured much of anything that was relevant.

Undertaking a quality initiative can be unsettling, at first. It is unsettling to go from the comfortable assurance that we are doing a good job, to a world in which no matter how good we get, the possibility of being better is always pressing upon us. Eventually, the organization comes to see that there is no final change we can make. And, there is no ultimate result we can achieve that will have our organization's quest for quality be completely realized. Neither prestigious new buildings, nor an array of high tech equipment, nor a prosperous net income can be a satisfactory substitute for continuous improvement in our products and services. It is like the song "Toyland," from *The Nutcracker Suite*. "Once you cross its borders *(into continuous quality),* you may never return again." Organizations such as yours and mine begin to discover new rewards and satisfactions in the continuous pursuit of quality, rather than in the comfort of keeping things on an even keel.

Why a process focus is necessary

I doubt if there is any company or institution that would say it isn't interested in improving quality. But continuous improvement doesn't happen just by talking about it. To integrate quality throughout an organization, we have to build an environment that encourages dynamism and change, not complacency and inertia. To instill the quality commitment in everyone, we have to recognize and reward those who are willing to try out new ideas—not those who are wed to the past. And to actually cause a cultural transformation, we must focus on the design, redesign, and management of our processes, not on our bottom-line results.

Clearly, this is not to say that a company doesn't need to have successful financial results to stay in business. But the organizations that have strong results over time don't accomplish that by watching the bottom line. Rather, their results flow from the quality of the goods and services offered. And quality flows from the processes that are used.

Connecting process work with customer focus, and not a cost reduction to improve the bottom line

When an organization's focus is on improving a process—any process—it means that our eye has to be on the customer of that process. And, keeping our eye on our customers is what a quality culture is all about. Look inside any company where process design and improvement is a fundamental value. You can be sure that service and respect for clients and customers is also a value. And, when clients and

Connecting process work with customer focus, and not a cost reduction to improve the bottom line
continued

customers are well served and respected, you can be sure that the company has something more at stake than simply a concern for the bottom line.

When SSM Health Care rolled out our CQI implementation, we were emphatic that we were not undertaking this effort merely to improve our financial statement. From our research, we of course knew that the idea of quality improvement had gained strength first in Japan, and then back in the U.S., because it reduced waste and therefore reduced costs and increased profitability. And, believe me, we had absolutely nothing against achieving those benefits. We were committed to achieving them in fact.

It had been startling to learn that 40% of U.S. health care costs, across the board, were attributable to waste. We said our system would no longer tolerate anything close to that level of wasted resources. All of the system's senior executives, commonly called the system management team, were confident that implementing CQI, with its emphasis on process design and problem solving, would result in the quality culture we wanted, as well as reduce waste and cost. But we wanted to make sure that everyone in our system fully understood this major effort for what it was—a commitment to quality, and not simply a cost-cutting project to improve the bottom line.

To avoid it being seen as a reduction program, we deliberately did not project any cost savings. Nor did we instill urgency, at first, about accomplishing our improvement projects within any set timeframe. We said that it might take five or more years to see real results. And, we made it clear we were not engaging in a pilot project or experiment.

If we had it to do over, I think we would have encouraged a greater sense of urgency in our projects to accomplish more in the early stages of CQI. We later learned, when we conducted a Breakthrough Series of improvement projects, that a lot of major problems can be solved within six months—and produce significant savings.

Benchmarking helped and we retained our principles

There was much that we learned also from visits and conversations with some other CQI organizations around the country. But even as we adapted and adopted various ideas, the one thing we did not adopt was the concern for the return on investment. Rather, the system management team was united in our stand that our efforts would focus on building a quality culture, whose number one principle is that patients and other customers are our first priority.

I want to dwell a little on how that principle fits into processes, their design, improvement, and management. In our human condition, you know that the tendency is always to blame a person or a group of people whenever a mistake is made or something doesn't work the way we think it should. But it is probably only about 15% of the time that breakdowns can be blamed on people. The other 85% stems from faulty processes or non-existent processes.

Processes are largely invisible in our cultures, and the result can be customer and worker dissatisfaction

Mostly, people don't see processes. We see tasks. One after another, the tasks line up like a disconnected series of things we do at work. If we are the one doing the tasks, we may set them out in a particular order to cross off our list. But at the end of the workday, we often wind up with a list of uncoordinated items that were either completed or carried over to the next day.

If we are on the receiving end of someone else's tasks, we may be left feeling fragmented. We call to check on a credit card bill, and the first person we talk with can't give us an answer to our question. We are put on hold for a few moments and another person comes on the line and asks how they can help. We ask our question all over again, and are told that the person who resolves those issues is in another office. They calmly and politely give us the number to call. Yet we are left feeling frustrated and even angry with the two people we've talked with. We are almost certainly going to be tempted to blame those two customer services representatives. Yet, it isn't the people who are to blame for the problem; it is the process, or more likely, the lack of one.

As a worker without a process, we may come to the end of our day's tasks lacking a sense of purpose or accomplishment. As a customer, we come to the end of a transaction, feeling irritated and asking the question, "Is anybody here paying attention?" A while ago I was at a department store in St. Louis. It was about 9:30 P.M., and I heard one salesclerk telling another one that her lunch break was scheduled for 9:45 P.M. The employee was complaining about having to go to lunch so soon after coming to work. The other employee agreed with her that it was stupid. As a customer overhearing their conversation, I quickly formed my own opinion about a store management that would be sending people to lunch at 9:45 in the morning! In very short order, the lack of a workable process left three people with negative opinions about how that store was run—all because of an unworkable or non-existent process. If you look at your own experience, you will recognize that it is an organization's processes that affect both how customers and employees think about it—whether positively or negatively.

Recently I saw a letter of complaint in the *St. Louis Post-Dispatch* from someone who had had a bad experience with a bill from a local hospital (not one of SSM's). The hospital has a national reputation for its medical treatment and has a number of top physicians on its staff. The letter writer blamed an "incompetent" billing clerk for the mistakes in her bill. I thought it was interesting that the customer was so sure that "incompetence" was the cause of the problem. That is usually where human beings tend to put the blame. Yet, from my experience with CQI, I bet the culprit was not a person, but a missing process! The clerk was left to appear inept in a situation over which he or she had no control. This was one of those "moments of truth" where the customer meets the company. And, as often happens, a person was blamed instead of a failure of process. The more complex and far reaching our businesses become, and the more sophisticated our products and services are, the more chances there will be for failures in those moments.

Even in a complex hospital setting, process design and improvement is both possible and necessary

There is perhaps nothing more complicated than the functions of a hospital that is open 24 hours a day, seven days a week. The endless loop of overlapping procedures, reporting structures, and lines of accountability within various departments make it seem impossible to design processes that work for our patients and other customers. And that was actually one of our concerns when we first considered the feasibility of bringing CQI into our system. But, it quickly became clear to us that there is no feasible alternative to process design and improvement for an organization committed to serving its customers. The company that tries to save itself the work that goes into process improvement is not saving anything. It is like the TV advertisement where the auto mechanic tells his customer, "You can pay me now, or you can pay me later." If an organization is not willing to invest the time, money, and effort to create an infrastructure for process design and improvement, it will pay over and over again—in wasted time, material, energy, and customer dissatisfaction.

Even before our system management team knew that process design and improvement was missing from our facilities, we knew there was something missing. In May of 1989, SSM Health Care was faring well in the health care industry. As I mentioned earlier, when we compared our facilities to others in our markets, we looked pretty good. Yet among our system management team there was a bit of discouragement and discontent. Throughout the system, there seemed to be too much satisfaction with the status quo. If something was working fairly well, our facilities, as a whole, tended to use ratings and rankings as indicators that we were doing as well as anyone else and better than most. Another source of discontent was that we didn't think we had the structures in place to make the best use of people's talents. And, while we had a strong mission and set of values, we had no way to operationalize those values throughout the system. When we implemented CQI, we gained the structure to help us integrate those values into everything we did. And, clearly, the process-focused character of CQI plays a big part in that.

Getting started in process design

The way the Baldrige National Quality Program describes a process is "linked activities with the purpose of producing a product or service for a customer within or outside of the organization." The very first lesson in designing or redesigning processes is learning to work with the people who represent the various activities that must be linked together; that is, the owners of the process.

Having operated for years in the management tradition of command and control, working on teams was not our forte. As you may know, in command and control structures, managers are trained from their earliest days to see themselves as the keepers of the knowledge. To be a good manager, of course, you would have to act like you knew everything about everything! When we honestly looked at the multitude of people and things we were accountable for, we saw there was no way we could know everything. In fact, it was a wonder that we accomplished anything when we pretended to know everything.

Lone Rangers and Mighty Mouses

Within the upper management of the system, we soon discovered that we divided into two groups: those Lone Rangers, who try to do everything solo. And the Mighty Mouses, imitating the cartoon figure that could fly in anytime to save the day. We quickly found out that process design is a team effort. To really work, every member has to know they have something to contribute and that they will be heard. If the team isn't representative, the process will be missing something.

This was illustrated dramatically by system management's first team effort when we began implementing CQI. Our project was deliberately a small one. We wanted to improve the cycle time for routing the mail among system management in the corporate office. In our work, we painstakingly designed a flowchart of what we thought the existing process was like. We posted the chart for all to see and asked for input. We received more than a dozen comments correcting aspects of our chart. The comments came from members of our support staff—none of whom had been included on our team!

Looking again at existing processes

A second important aspect in process design is being willing to keep looking at the existing process, and asking, "Why is this step here, and why is that step there?" When we do this, we ultimately get to the root cause of the breakdown in the process. But it takes patience. In a culture in which people think they already know how something works, it is harder to look at something simply and objectively. When we do get to the root cause, nine times out of ten we don't find an uncaring employee; we see a non-existent or poorly designed process.

In process design, it is usually clear that the original process, or what was passing itself off as a process, is so flawed that all there is to do is to start from scratch. In that case, the team would look at what it is that happens—what or who must get from point A to point B—and say, "Under the ideal circumstances, what would the design of this process be?"

Addressing the willingness to change

In the hospital setting, one of the things we saw is that getting health care professionals to create a standard process for something is about as natural as herding cats. Well-trained professionals, who are confident of their skill and proud of their abilities, simply do a job. They are not looking for a manual. One surgeon might come in and ask the surgical nurse to set up the instrument tray for an appendectomy in a specific way. He is sure that his is the most effective way to have the tray set up. Another surgeon will ask for a completely different set up and so on.

What motivates physicians and nurses to improve processes is evidence that there are wide variations in outcomes based upon the processes used. When that evidence is collected and presented, a team has its case for process design or redesign. Often you don't even need evidence to make the case for a process redesign. Many people can see that the process is flawed.

How many times as a customer yourself, have you complained to the person serving you that something just doesn't work? And how many times have you heard the employee say, "Well, that's just the way we are told to do it. " Or, "You'll have to

Addressing the willingness to change,
continued

talk with my manager, because that's our policy." How can people in an organization take pride in their work, if they know that what they are being asked to do doesn't serve customers, but they have to keep doing it anyway? Whenever I see a situation like this, it reminds me Rita Mae Brown's definition of insanity. She said, "Insanity is doing the same thing over and over again, expecting a different result."

An example

In an organization that has an infrastructure to support the work of process design and improvement, people are empowered. They feel like they have a stake in how things operate, and they have a say in serving the customer well. We had a situation with one of our hospitals and nursing homes that were located fairly close to each other. Typically, when an elderly patient had recovered sufficiently to leave the hospital, they would go to the nursing home. But whenever a doctor happened to discharge a patient on a Friday morning, the patient would arrive at the nursing home when the weekend schedule was in effect. If the doctor did not write the orders for the patient, or could not be reached by phone, the patient would be left without a physical therapy plan of care or medication until Monday. Patients would sometimes be left waiting until Tuesday before they could have their first physical therapy session, and thereby miss out on one or two days of therapy.

Two teams, one from each facility, got together to create an improved process. The plan gives each discharged patient a standard set of orders that accompany him or her to the nursing home. Now, when a patient comes to the nursing home, no matter what day of the week, he or she has a complete set of orders for pain medications, any other medications, and a physical therapy schedule, so they can begin their therapy at the earliest possible time. This was simply a matter of designing a discharge planning process from the point of view of our customer, not from the point of view of the care providers. The design team first saw that real people had to be transported from point A to point B. They asked the question, "In an ideal world, what would be the best process for moving him or her to that point?"

Another example

In a similar occurrence, one of our hospitals has altered the entire process of its care for the dying by imagining themselves in the place of the patient and the patient's family. Previously, the hospital personnel had simply set out to ensure that the dying patient was well cared for while in the hospital. But when a team of caregivers looked from the perspective of the patient and the patient's family, they saw that many things have to be provided when someone is dying, and not all of them are within the hospital. The team created a process that includes several other sectors of the community—churches, home health agencies, funeral homes, grief support groups, and so on. Sometimes process design rightly expands beyond the service or product you offer. Companies might legitimately take the view that they can only be responsible for their product or service. But I don't believe an organization that is seeking to transform its culture with process design and management can put limits on itself.

Collaborative agreements to serve customers better

Notice all of the collaborative arrangements that are springing up among hotels, airlines, car rental agencies, credit card companies, and membership organizations like AAA and AARP. Integrating the services and products of several providers is an example of process design that comes from standing in the shoes of the customer, rather than stopping at the boundaries of the corporation.

Eliminate hassles

Philip Crosby, one of the leaders of the quality movement in this country, has equated quality with "hassle elimination." Just ask your customers or clients for their opinion of your product or service and you will understand how quality and freedom from hassles can seem synonymous. Customers tend to be delighted when processes are simple, convenient, logical, and hassle-free. And, they tend to be frustrated, angry, irritated, and impatient when they are not.

> *Customers tend to be delighted when processes are simple, convenient, logical, and hassle-free. And, they tend to be frustrated, angry, irritated, and impatient when they are not.*

Example

One of our facilities discovered that its phone answering process was an annoyance for callers. In tracking the calls to the switchboard, it was found that 100 people a week who were put on hold, eventually got tired of waiting and hung up. The solution was not to increase the number of people answering the phone, but to change the way the phone system was programmed. The new system provided callers with more information and allowed them to leave a phone number for a call back. In six months, the number of hang-ups was reduced to 10 per week. Designing a process that respects and serves the customers of that process—whether they are patients, clients, or employees—requires paying attention to the customers' interests.

Your culture is a process, too

Creating a culture of continuous improvement means that process improvement teams eventually will form on their own when employees know that is expected and encouraged. A quality culture calls all of its employees and managers to be acute observers of what happens and how things work in their areas. This requires employee education, involvement, and development. The people who do the job are the ones who have the best vantage point on how it can be improved. But they have to be trained in how to go about improving something in a way that makes a difference.

Our CQI classes provide valuable learning on how to get to the root cause of variations, how to measure variations, and how to determine if something is a common cause variation or a special cause variation. This knowledge means the difference between employees who can implement strong improvements or those who just tinker with processes.

The value of measurement

In all of the work done on process design and management, measurement becomes a critical factor. For those who aren't statistically oriented, measuring the process may at first be simply a pain in the neck. Yet, without measurement, supervisors and managers are left to the same opinions, suspicions, hunches, and assumptions that have traditionally driven our management decisions. The subjectivity of decision making without measurement often turns a decision into guesswork.

One of the five CQI principles we adopted when we began implementation was termed "decision making by objective data." I'm sure that in your industry and in your particular company, you have long had certain measures that you keep your eye on. You have to have objective data that you follow perhaps over many years, to track trends in quarterly sales, or to notice the seasonal availability and price of materials or ingredients. In health care, we have long used such measures as occupancy rates, lengths of stay, and the ratio of Medicare/Medicaid patients to privately insured patients and charity cases. If you managed anything in a hospital, you knew the numbers to keep your eye on. Regardless of the industry, there are always the standard measures that give you a view of reality. But when it comes to quality, the traditional measures don't tell us much. Our length-of-stay numbers gave us an insight into expected occupancy, or income projections, but they do not tell us anything at all about customer satisfaction. They do not give us access to ways to improve a process. If patient satisfaction is affected by the length of the wait times in the emergency department, then we had better find a way to measure how long people were waiting.

There are many federal, state, and local bodies that regulate hospitals. Their interest is in compliance with policies and regulations, not customer satisfaction. The measurement of processes gives us a window into the customer's experience, which, when it comes to quality, has to be at least as important as compliance.

In his book *The Fifth Discipline*, Peter Senge wrote about the difficulty of managing quality in a service business due to the intangibility of service activities. He said there is a strong tendency to manage service businesses by focusing on what is most tangible—numbers of customers served, costs of providing the service, and revenues generated. "But," Senge points out, "focusing on what's easily measured leads to 'looking good without being good,' to having measurable performance indicators that are acceptable, yet not providing quality service."

How can we know what aspects of a process are wasteful and inefficient if we are not measuring anything? How can we tell what our customers' experience of our service is, if we don't have mechanisms for gathering and analyzing their opinions? Whether it is phone rings, trips to the laboratory, lab specimen handoffs, or the number of minutes it takes to clean and prepare an operating room, measuring things gives people data to make decisions. For example, if a quality team reduced the turnaround time in an operating room from seven to three minutes and all of a sudden, that number started rising again, the team would have to go back and look at the steps in the process. The team would collect data to see what was really going on.

The value of measurement, continued

In a previous life, when I was an operating room supervisor, I always knew that surgeries started late because anesthesia didn't get there on time. But the fact is, that was only my guess; I never collected any data. It may have been true, but I really didn't know. If we don't know the facts about a process, then our attempts to impact it for the better are going to be only guesses. Yes, sometimes we may get lucky and guess correctly, but more often than not, we will not.

Trust and cynicism

One of the things we saw early in our implementation of CQI is that human beings' natural cynicism and lack of trust can stop the development of a quality culture. The system management team and I discovered early that we could not put a series of demands on people's time and attention and then just watch to see what happened. We had to be in there learning and teaching ourselves. We had to be on teams. We had to go through the process of getting to root causes in the functions of our corporate office. We had to learn how to measure and to use the tools of process analysis. People in your organization have to see that you are walking the talk.

Process design includes system thinking

The leaders of quality organizations also have to be up to managing the entire quality improvement effort over the long term. Process design/redesign and improvement do not succeed without consistent management. Nothing stays in existence by itself. The tendency, in fact, is for everything to go out of existence.

Knowing your purpose and sticking to it

One of our system's greatest compliments came from Dr. Donald Berwick, the nationally recognized health care quality expert. He acknowledged our "constancy of purpose," in continuing to manage CQI in our system over a long period of time, through thick and thin, even when it gets tough. As you know, "constancy of purpose," is W. Edwards Deming's first principle of quality. With the perspective of time, those of us at SSM Health Care can see now why that principle is so important. Without management's commitment to keeping the quality effort in motion, it will surely stop.

Process management

Sometimes we see habitual patterns or behaviors going on for long periods of time. But if you observe your organization, I think you will see that the best behaviors quickly lose their edge if managers stop paying attention. People may keep doing the same things for a while, but whatever it was that lit people up would disappear without management, without motivation, without acknowledgement.

If the supervisors, managers, and

> *If you observe your organization, I think you will see that the best behaviors quickly lose their edge if managers stop paying attention. People may keep doing the same things for a while, but whatever it was that lit people up would disappear without management, without motivation, without acknowledgement.*

Process management,
continued

executives let down the attention they pay to those processes, you will see the employees letting down as well. The *word*—implicitly or explicitly—gets around that those ways of doing things are not quite as important as they once were. Then you wind up with a process that is not being managed, or we could say, a process that is not being maintained. Until it starts to become second nature to improve processes, it will require managers continuously generating the environment in which improvements can be initiated. Otherwise the people closest to the work and those in supervisory positions, will tend to say they don't have time to get into it.

Inconsistent management of process improvement can also be a problem. If employees in a company or institution sense that processes are important to some managers and not to others, this sets up confusion, uncertainty, and frustration. Do I or don't I follow this process? Is it important or not? People get mixed signals about their jobs and what is important. It leads to cynicism. Quality service and customer care and respect may be perceived only as window dressing that is spruced up when some of kind of quality inspection is going on. This does not support a transformed culture—it is more like business as usual.

Process improvement

Encouraging and requesting that employees design and improve processes also requires openness with the company's or institution's business. We found executives at some of our facilities were reluctant to divulge certain information that teams needed to analyze processes. There is a risk in trusting people with key information. Employees, who come to know processes intimately, can be tempted to use the access to money or data in a fraudulent way. We've had that occur recently in our system. A finance officer in one of our smaller corporations devised a way to embezzle over a million dollars over a period of several months before anyone noticed.

The three people who had direct responsibility for that area came to my office, embarrassed and devastated. They offered to resign. I didn't accept their resignations. Instead, I asked them to use that experience to determine where the process broke down. I asked what happened in the process and how could it be improved so that this activity couldn't happen again. Sometimes, the amount of money involved might be a temptation to give up on the process. This was a significant amount of money. But what we had to do was find the breakdown and improve the process. What they found was that there was in fact a process in place, but it wasn't being followed.

The importance of trust

In a quality culture, the other temptation is to identify where people have authority and say, "I'm taking that away." But, if we are truly a quality improvement culture, committed to process improvement, we cannot afford to stop trusting people. We cannot afford to withhold necessary information. We have to design processes to keep temptation out of people's way. And then make sure we follow the processes. Process management also means that processes are subject to periodic review—to see if they are really working as intended, even if the variation is in control.

Achieving organizational transformation

In my sharing with you today, I have told you some of the things we have learned about building a quality culture through process design, improvement, and management. But there is one last and most important thing about this topic of processes. There is something truly powerful that happens in an organization and among employees when they are at work on process improvement. Looking for ways to improve the way we do things for our customers and clients alters the vantage point from which we see. And by altering our perspective, we transform what we have been dealing with—even if we have been dealing with it for years.

> *There is something truly powerful that happens in an organization and among employees when they are at work on process improvement. Looking for ways to improve the way we do things for our customers and clients alters the vantage point from which we see. And by altering our perspective, we transform what we have been dealing with—even if we have been dealing with it for years.*

Give the power to improve processes to your managers and employees and watch what happens. Instead of an automatic, unthinking approach to a routine, people start paying attention to what they are doing. They start standing in the place of the customer and discovering how the process or lack of process is being experienced over there, rather than simply performing a task. The human spirit comes alive in people who know they have the capacity and the authority to make a difference in their work.

The job of the organization is to take the spirit that abides in every one of the people who work with us and make that a part of the organization's own transformation. In SSM Health Care, we say that the spirit of people that arises when they are engaged in improving the processes of their work is one of the essential components of our institution's transformation. It seems an amazing phenomenon—but it's nevertheless true. The act of designing, redesigning, and continuously improving processes, alters the spirit of the people who authentically engage in that work. And, the transformation of the whole organization comes about as more and more people who work there feel free to express that spirit.

> *The job of the organization is to take the spirit that abides in every one of the people who work with us, and make that a part of the organization's own transformation.*

When I speak of this spirit, I am not speaking of it as a religious concept. The spirit is what is at the core of your being. It is what makes you the person you are; it is that ineffable quality that is present when people are at their best. We see it throughout SSM Health Care, in people of all cultural, ethnic, and

Achieving organizational transformation, continued

social backgrounds. And you see it, too, in your organizations. We see it in the hospital cafeteria worker who stands behind the steam table throughout her shift and has a friendly smile for each person who goes through the line.

We see it in the housekeeping supervisor who goes out of his way to personally greet every person on his crew and expresses concern about how each one is doing. We see it in the surgeon who, after 12 straight hours in the operating room, decides to make one more stop on the surgical floor to see how a patient is feeling. We see it in the social worker that remembers a special need of a patient about to be discharged and makes an extra phone call to accommodate that need. The spirit of a company emerges from the spirit of the people who work there. The individuals fuel the organization with their spirit and they in turn find a place to express their desire to serve and contribute and make a difference with others.

Very early in our work on quality, I received a letter from a security guard who worked on the pay parking lot at one of our hospitals. She told me that she had recently allowed one of the visitors to the hospital to leave the parking lot without paying. The visitor had gone to the hospital to see her husband and did not realize until she got to the exit gate that she had left her purse at home. The security guard waved her through the gate and wished her a good day. In her letter to me, the employee said that the day before she had heard me give a talk about putting the customer first, and she decided that this was an opportunity to do so. Her spirit, her desire to be generous was now freed, rather than constrained, by her work.

An unhappy employee will not deliver the compassionate care we say they will deliver.

> *Employee satisfaction is the gateway to customer satisfaction. You cannot have people thoughtfully and sometimes painstakingly creating processes from the vantage point of the customer, and not see their own joy in their work come alive. You cannot have people thinking from the customer's perspective to create processes, and not have their sense of compassion expanded. You cannot have people looking at how to make a process better, and not have employees start expressing their creativity.*

Employees need to experience the same level of compassion from us that we are expecting them to pass on to others. Employee satisfaction is the gateway to customer satisfaction. You cannot have people thoughtfully and sometimes painstakingly creating processes from the vantage point of the customer and not see their own joy in their work come alive. You cannot have people thinking from the customer's perspective to create processes, and not have their sense of compassion expanded. You cannot have people looking at how to make a process better, and not have employees start expressing their creativity. You cannot have teams working to redesign a process so that customers will be delighted, and not have that way of working cause a new spirit in the culture of the entire organization.

Conclusion: the energy of people becomes infused with the mission

When the people of an organization have the opportunity to really be alive, vital, and creative in their work, that life and energy is infused into the organization. We say that the opportunity to design and improve every process throughout an institution is how people get to bring that life to their work. The organization's mission starts being fulfilled, not on a static piece of paper, but in the actions and words of its people. The mission, the very purpose of the organization, is fulfilled through people. And people begin to have more fulfillment in their lives because their work is truly about serving people. Believe me when I say, if your organization and mine are designing processes that honor and respect our customers' needs and wants, then we are not designing processes that will merely improve our bottom line, or even only improve our products. We are designing processes that literally transform our organizations and the people in them. When that happens, we do not have to be concerned about how to make our organization successful. When that happens, our organization is successful. It is fulfilling its mission and it therefore cannot help but flourish in every sense of the word.

Author information

Sr. Mary Jean Ryan is president and CEO of SSM Health Care, headquartered in St. Louis, Missouri. It is one of the largest catholic health systems in the United States. She has been a member of the Franciscan Sisters of Mary for more than 35 years, and was appointed the first president of the restructured SSM Health Care System in July of 1986.

Sr. Mary Jean has emphasized three key themes during her 11-year leadership: commitment to Continuous Quality Improvement, preservation of the earth's resources, and enhancing ethnic and gender diversity.

She has a bachelor's degree in nursing from St. Louis University, and a Master's in Hospital & Health Administration from Xavier University.

Sr. Mary Jean has received a number of awards over the years, including The Brotherhood/Sisterhood Award from the National Conference of Christians and Jews in 1990, The Missouri Governor's Quality Leadership Award in 1997, and The Distinguished Health Care Ministry Award from Archbishop John May in 1998.

In 1999, SSM Health Care recieved the Missouri Quality Award, and it was the only health care system in the nation to score high enough to receive a Baldrige site visit.

Editorial support for this article was provided by Laurence Smith.

Living and Breathing a Customer-Centered Culture

Authors

Christine Kelly, M.A., M.T. (ASCP), CLS, Director of Laboratory, HealthSystem Minnesota, Minneapolis, Minnesota

Elizabeth Lentz, B.S., M.T. (ASCP), Regional Laboratory Manager, Park Nicollet Clinic, HealthSystem Minnesota, Minneapolis, Minnesota

Brief background on Park Nicollet Clinic

The Park Nicollet Clinic, HealthSystem Minnesota, is a 400 physician multi-specialty group practice with 19 individual clinics located in Minneapolis and throughout its suburbs. In 1993 Park Nicollet Clinic merged with Methodist Hospital HealthSystem Minnesota, a 426-bed facility located a half–mile from the clinic's main campus. HealthSystem Minnesota also includes the Primary Physical Network, consisting of 9 clinics and 36 physicians, the Institute for Research and Education, the Foundation (a fund raising entity), and all of the 1,000 physicians with admitting privileges to Methodist Hospital.

The 17 Park Nicollet Clinic Laboratories and the Methodist Hospital Laboratory have also recently merged. The Laboratory system now consists of over 220 technical and nontechnical personnel who collect and perform diagnostic testing on patient specimens. The Quality Improvement initiatives of the clinic and hospital labs are just now beginning to be integrated. This article describes the activities that have occurred in the Park Nicollet Clinic Lab over the last seven years.

The Quality Improvement journey

Park Nicollet Laboratory has been on a journey of Quality Improvement for the last seven years. We have been continuously learning about Dr. Deming's principles and trying to incorporate them into our daily work processes. A major challenge in trying to integrate our efforts consistently throughout the entire laboratory system is the fact that many clinic lab sites are so geographically diverse.

Traditional changes, then a move toward a real transformation

Park Nicollet began its Quality journey in December of 1989 when a group of top clinic leaders participated in a formal week-long seminar. Afterward, a half-day training session was developed to introduce Quality Improvement to the rest of the clinic staff. However, teams were formed primarily as a reactive response to variation and special causes in the system. Many members of our staff viewed these meetings and the new talk about Quality Improvement as something external to the daily process, draining their energy from already busy days.

Traditional changes, then a move toward a real transformation, continued

In 1993 the organization started to move away from a traditional approach to Quality Improvement to one of "transformation." In a transformed organization, there is continuous learning about Quality Improvement through application in daily work processes rather than separate, disconnected training sessions. There is a global understanding of the organization's systems and processes, as well as the capabilities of the current systems. The organization identifies key strategic goals that are essential to its survival. An example is: "Providing service so good that our patients refuse to go elsewhere." Steps are taken to proactively build processes that support the goals and ensure they are attained. Key measurements are taken of the daily work processes so process corrections can be implemented to stay the course. With transformation, everyone focuses on total quality in every aspect of their jobs. (Figure 1)

Figure 1. A Timeline from Continuous Quality Improvement to Transformation

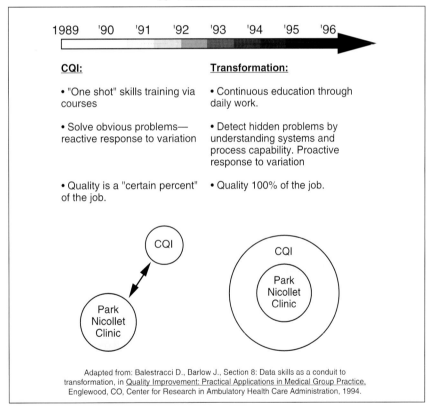

1989 '90 '91 '92 '93 '94 '95 '96

CQI:

• "One shot" skills training via courses

• Solve obvious problems— reactive response to variation

• Quality is a "certain percent" of the job.

Transformation:

• Continuous education through daily work.

• Detect hidden problems by understanding systems and process capability. Proactive response to variation

• Quality 100% of the job.

CQI

Park Nicollet Clinic

CQI

Park Nicollet Clinic

Adapted from: Balestracci D., Barlow J., Section 8: Data skills as a conduit to transformation, in Quality Improvement: Practical Applications in Medical Group Practice, Englewood, CO, Center for Research in Ambulatory Health Care Administration, 1994.

A timeline for Quality Improvement

The Laboratory has paralleled the learning curve of the rest of the clinic, and this is shown in Figure 2. The italicized items indicate education activities we have participated in, or informally designed ourselves, to move our lab team along our own learning curve. The items in bold text are different teams that we started. The first teams, UA TAT (Urine Analysis Turnaround Time), CHEM TAT (Chemistry Turnaround Time) and MICRO Team (Microbiology Team) were teams formed as a "reaction" to

A timeline for Quality Improvement, continued

Figure 2. Park Nicollet Clinic Lab Quality Improvement Continuum

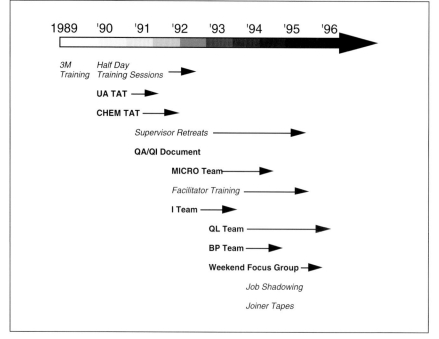

customer/physician service complaints about test turnaround times. Later, in 1993 when we began to understand transformation, teams like the I (Integration) Team and the BP (Best Process) Team were formed to proactively look at our processes and make improvements before receiving customer complaints.

The Quality Leadership Team made the biggest impact

As we began to create a cultural transformation in July of 1993, all lab staff needed to understand Quality Improvement Principles and make them a part of the laboratory culture. It had to become the way we go about doing our daily work. The Quality Leadership Team (QLT) was formed, in the hopes that this group of people would become the critical mass needed to move our efforts forward. The activities of the QLT have been some of our most exciting accomplishments.

The QLT consisted of 18 people, which is larger than normally recommended for teams. However, we felt strongly that all lab supervisors needed to participate because, as Dr. Deming taught, the commitment must first come from management. We wanted to balance the team with an equal number of front-line staff. Everyone on the team is considered an equal; there are no titles.

We meet for two hours each month. The first hour is an education piece, such as viewing a videotape from the Deming tape library, followed by discussion on relating the theory to our laboratory. The second hour is spent working on the current initiative. We have also used a subcommittee format to get things done between the monthly meetings. The QLT has had the biggest impact on the laboratory, and has become the foundation of all our Quality Improvement efforts.

The first initiative: Redesign of the old Quality of Service Evaluation (QSE) system

The first QLT initiative was to revise the Quality of Service Evaluation (QSE), a documentation form for problems and errors that had been in place for many years. It was perceived by staff as a very punitive system. For instance, the first question at the top of the form was, "Who was involved in this incident?" It also promoted system changes for special causes. Staff also viewed the forms as going into a black hole. Once completed, they were turned into the supervisor and reviewed by the lab management team. However, there were no mechanisms for reporting back to staff. Needless to say, there was very low compliance in the use of the QSE form.

Opportunity statement

It became apparent early in our discussions that we needed to broaden our focus beyond just the problems of the QSE form. We decided to create a comprehensive foundation for all Quality Improvement efforts. Our opportunity statement was to design a process that would include promoting opportunities for staff involvement, educating them on the process, redesigning those tools to document Quality Improvement activities, and building in a clear mechanism for following through and providing feedback to staff (Figure 3).

Figure 3. Quality Leadership Team Opportunity Statement

To design a process for laboratory personnel to identify problems and improve systems.

The process we design will include the following:
- Promote opportunities for staff involvement in the improvement process
- Staff education on the improvement process
- Redesigned tools
- A clear mechanism for feedback and follow-through.

Surveying staff, gaining feedback and identifying gaps

First, the QLT surveyed the staff on the problems with QSE system, from which we received about 10 pages of additional negative comments. Then, we flowcharted the current process. The flowchart was very complicated, which illustrated why the QSE was ineffective. The gaps in the system were identified from the surveys and flowchart and categorized into five groups:
- Education/Responsibilities
- Tools
- Timeliness
- Outcomes/Closure
- Universal Application.

Forming subcommittees

The list of system gaps that needed to be addressed was quite extensive. Therefore, we decided to form three subcommittees from the QLT members. The Procedure subcommittee's charge was to design a new process for documenting problems and errors. The Tools subcommittee would then design a documentation tool or form related to the new process. Because these two issues overlapped, the Procedure subcommittee members actually continued on to become the Tools group. The Education subcommittee was responsible for developing a training session that would teach all lab staff about the new QSE process and instruct them on how to complete the new documentation form.

Systems Process Improvements Flowcharts

The Procedure subcommittee identified three general categories of quality improvements that can occur: (1) Suggestions for Process Improvements; (2) Problems of Immediate and/or Important Nature; and (3) Recurring Problems. For each category, a flowchart was designed that outlined the steps an employee would take to resolve a problem or bring forward an improvement idea. A new documentation tool called a Systems Process Improvements Form (SPIF) was also designed related to each category. The SPIF has the flowchart right on it to make it easy for the employee to follow each of the steps.

Employees write their suggestion or state the problem on the SPIF form and discusses it with their supervisor. The supervisor plays a key mentoring role and ensures that the employee actively participates in the improvement process. However, a copy of the SPIF is also sent to the QLT who keeps track of what is happening globally throughout the lab system and also ensures that no suggestions get undermined due to lack of support from a supervisor. There is a mechanism to report back to the initiator so they understand if there is a good reason an idea is not implemented. Once all the flowchart steps are accomplished, the completed SPIF form is again forwarded to the QLT which collates all SPIF activity and reports it back to all staff.

SPIF 1: "Suggestion for Process Improvement"

The "Suggestion for Process Improvement" flowchart (SPIF 1), shown in Figure 4 on the following page, outlines the steps to be taken when an employee has an improvement idea. Everyone in the lab is encouraged to identify ways they can do the job better, simplify systems and improve the quality of lab services we provide by asking questions such as, "Why do we do it this way?" "Wouldn't it be better if...?" or simply "I don't have any ideas to make it better, but I know that what we're doing just doesn't work!" A goal is to facilitate process improvements as quickly as possible. So, this flowchart has two pathways. One accommodates quick, "no brainer" ideas that everyone quickly agrees on. The other uses the Joiner 7 Step Method for more detailed planning of the improvement.

SPIF 1: "Suggestion for Process Improvement," continued

Figure 4. The "Suggestion for Improvement" Flowchart

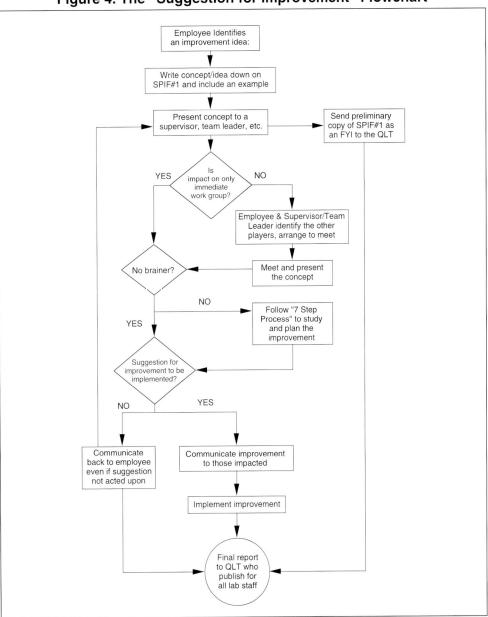

SPIF 2: "Problem with Immediate and/or Important Nature"

The "Problem with Immediate and/or Important Nature" flowchart (SPIF 2), shown in Figure 5, is used when a problem/issue occurs that needs an immediate resolution in order to provide appropriate customer service. For instance, perhaps a patient is unhappy because he felt he had to wait too long to get his blood drawn. Another example is that we collected a patient blood specimen in a wrong container and we need to correct it immediately because the physician is going to be looking for the test results right away. Documenting problems of this type is similar to the previous QSE process. These problems are often one time "special causes," but are monitored over time to determine if there is a recurring problem requiring a change in the system.

SPIF 2: "Problem with Immediate and/or Important Nature," continued

Figure 5. The "Problem with Immediate and/or Important Nature" Flowchart

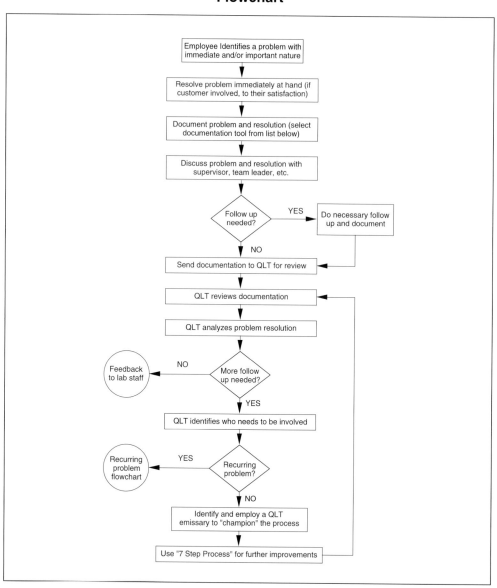

SPIF 3: "Recurring Problems"

The "Recurring Problem" flowchart (SPIF 3), shown in Figure 6, is a way to handle problems that seem to never go away. For instance, someone might say, "It seems like this is the third time this week I've received an unsatisfactory specimen from a medical department." Or, the QLT may notice a pattern of the same type of problem that occurs globally throughout the lab system. For instance, twice a week a particular lab test is ordered incorrectly by the nurses and it happens at each clinic site. Initially, that might not seem to be significant from the viewpoint of one site. However, when you multiply that error across all 19 clinic sites, it becomes a bigger system-wide problem. This flowchart always uses the Joiner 7 Step Method because if a problem has been recurring, it warrants an in-depth approach to solving it.

SPIF 3: "Recurring Problems," continued

Figure 6. The "Recurring Problem" Flowchart

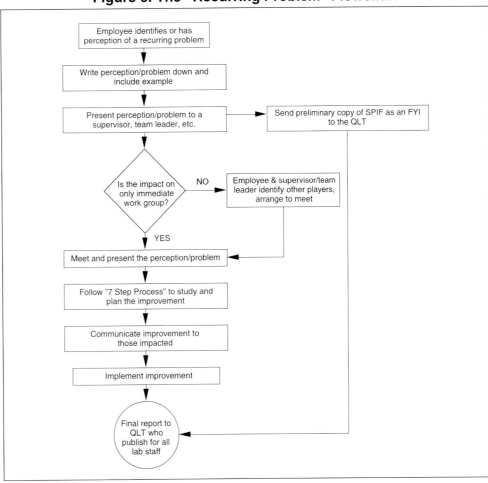

Developing a training program

Once the Procedures and Tools subcommittee defined the three flowcharts and accompanying SPIFs, the Education subcommittee developed a 1 1/2 hour training session. The purpose of this was to review some basic concepts of Quality Improvement, explain the three System and Process Improvement flowcharts that were developed and have the staff learn how to complete each of the three SPIFs. The QLT members were the trainers. We broke up into teams of two trainers and all 132 people throughout the lab sites attended one of the sessions.

We started the session by emphasizing that quality improvement is an ongoing process. We cited examples of various lab teams and improvements that had already been accomplished. We told people who Dr. Deming was and spent time talking about the 14 points—we related them to real life examples within the laboratory and tried to categorize them as relating to "everyone tries to do their best" and "look at systems, don't blame people."

Next, we introduced the Joiner 7 Step Method. We role-played a teenager learning how to drive and look ahead down the road. This was to emphasize that we want to be proactive and look ahead to spot potential roadblocks that prevent us from

Developing a training program, continued

doing a good job. The 7 Step Method helps us move away from reactive putting-out fires, to implementing improvements that really work.

The rest of the training session was spent explaining each of the flowcharts and using some real laboratory problems to practice filling out the SPIFs. We now repeat this training session on a quarterly basis for all new employees so they are indoctrinated into this culture very early on.

Feedback to staff and follow-through on SPIFs

One of the goals in the Opportunity Statement was to provide a clear mechanism for feedback and follow through of the SPIFs. The QLT members also serve on two subcommittees that meet outside of the monthly QLT meeting. The "A-Team" is responsible for monitoring the SPIF 1s. The "B-LTs Team" collates the SPIF 2s and SPIF 3s. Each SPIF is assigned a QLT member as a facilitator to mentor the person who initiated the SPIF through each of the flowchart steps and ensure that the SPIF makes continuous progress until completion. The A-Team and B-LT Team keep track of all the SPIFs on a spreadsheet which lists the following:

- Origination date of the SPIF
- The improvement suggestion or problem that occurred
- QLT member assigned as facilitator
- Progress on what's been done to date
- Further action items identified
- Mechanism to be used to report outcome to staff (e-mail, meeting, etc.)
- Date of closure for the particular SPIF
- Any future follow up actions that are needed.

Communicating results

We publish a newsletter, the Quality Connection, to inform staff about the outcomes of the SPIFs and eliminate the "black hole." We hope the newsletter also reinforces Quality Improvement as a part of the daily lab culture. The Quality Connection includes an education section and columns written by both the A-Team and B-LTs Team highlighting process improvements that were implemented. We attach a copy of the SPIF spreadsheets described above so the staff can see everything that is going on and hopefully be encouraged to think of additional ideas.

Celebrating successes

The QLT also organized the first annual "Celebrate Our Success" evening. It included a light dinner, and various staff members made presentations of the SPIFs they initiated, the process they took to implement changes, and the final outcomes. The evening was another way to eliminate the "black hole" for feedback, and an opportunity to pat ourselves on the back for improving laboratory services.

Examples of changes made from this process

This article will now describe three examples of changes that were made because of suggestions made by front-line employees using the SPIF process. These examples help demonstrate the buy-in of the staff and the transformation in their thinking. The examples illustrate eliminating rework, understanding systems and improved customer service.

1. Eliminating rework

Because we are geographically diverse, there are significant differences between the satellite labs and the central lab. In order for employees to better understand the processes at other locations, a program of job shadowing was initiated. Staff from the central lab can spend a day at a satellite lab and vice versa.

It was while job shadowing that a satellite lab employee noticed that there was duplication of work in the processing of 24-hour urines (See figure 7).

A 24-hour urine collection results in a large volume of liquid. At the satellite lab, the total volume is measured and then a small portion (aliquot) is transferred to a test tube that is sent to the central lab. In the processing area of the central lab, the specimen was transferred again to a smaller test tube in order to continue processing.

There did not appear to be a good reason why the specimen could not be put into the correct size test tube at the satellite lab. A one month pilot program was tried and there were no problems. The change was then implemented throughout the system. The specimen transport procedure was revised and the change was communicated to all staff. Estimated cost savings for a year is a little over $1,100 in supplies and labor.

Figure 7. Specimen Handling

Issues:
- Satellite lab employee noticed rework in handling of specimens while job shadowing in specimen processing
- Eliminating interim container would result in less handling & expense.

Action:
- Procedure change developed
- No problems encountered during one-month pilot
- Recommendation made to use throughout lab system.

Outcome:
- Specimen transport procedure revised
- Change communicated to all staff
- Cost savings of $1,130 per year in supplies and labor.

2. Understanding systems

When only a small amount of blood is needed for a laboratory test, we use a micro-specimen collection device. The staff was not satisfied with the one we were using. They felt that there must be something better, but no one had an answer as to what that "something" was (see Figure 8).

2. Understanding systems, continued

There were several problems with the collection device. It appeared that many of the specimens we collected could not be analyzed and had to be rejected. Calling a patient and asking them to come back and have their blood drawn again is very unpleasant for everyone involved.

Secondly, the devices were difficult to mix. They had to be mixed with a snapping motion and after employees had done that several times a day, many of them complained of wrist pain.

An interdepartmental team was formed, and they used the Joiner 7-Step Method to investigate several different collection devices. The team selected one because of its ease of collection and mixing to be tested by several sites. Test results from both the old and proposed device were statistically evaluated to determine any variation. The technicians evaluated the new device on its ease of draw, ease of mixing and number of rejected specimens, and then submitted comments on which device they preferred.

The benefits have been numerous. Once they were trained on how to use the new, different collection device, the employees liked it. Since it is less costly than the previous one, we anticipate a savings of about $813 per year. There is also the potential savings from preventing employee injury and the benefits of better customer service. Recalling patients is costly in time, supplies, and especially customer satisfaction.

Figure 8. Micro-Specimen Collection Device

Issues:
- Staff dissatisfied with current micro-specimen collection device
- Many specimens are rejected
- Difficulty in mixing has resulted in wrist pain for some employees.

Action:
- Interdepartmental team formed
- "Joiner 7 Step Method" used
- New, less expensive device selected and tested at selected sites
- Proposed collection device met criteria.

Outcome:
- Staff trained in using new device
- Procedure revised
- Supply cost savings of $813 per year
- Potential staff injury prevented
- Fewer patients recalled because of specimen rejection.

3. Customer service

A new employee noticed that we seemed to draw a lot of blood on all of our patients (See Figure 9). Patients had also made that comment, but we did not pay too much attention to them. Since she generated a SPIF, an investigation was done to determine if we were collecting more blood than we really needed.

The first thing that was discovered was that our Laboratory Information System (LIS) was programmed to generate one label per test. Therefore, whoever was drawing the blood was usually drawing one tube of blood for each test. The first action taken was

3. Customer service,
continued

to reprogram the LIS so that there would be more tests per label.

Next, key players from each laboratory department met to determine the best way to collect blood. A balance had to be carefully maintained between "too much" and "not enough." The problem with "not enough," of course, is that the patients would have to be recalled or some of the tests that the physician ordered could not be performed.

The process of determining what tests can be grouped on one label is an ongoing process and is reviewed on a periodic basis. A change or addition in test methodologies or changes in patient guidelines results in a reevaluation of the test grouping to see if new or different combinations need to be used.

The amount of blood required has decreased, and the implementation of a process to evaluate the amount of blood drawn fulfilled a requirement for the lab's regulatory accreditation.

Figure 9. Volume of Blood Drawn

<u>Issues:</u>
- Currently one label generated per test
- Number of tests per label programmed into system
- Need more tests per label.

<u>Action:</u>
- Meeting arranged with key players
- Preliminary change to more tests per label implemented
- Project ongoing with additional tests being evaluated.

<u>Outcome:</u>
- Amount of specimen required has decreased
- Implementation of process fulfilled a requirement for the laboratory's accreditation.

Next steps: Shifting to a Customer Centered Focus

Once we felt that we had a good system in place, it was time to move on to the next level and shift to a Customer Centered Focus. Our employees knew how the system worked and how to improve processes that weren't working. Everybody was very enthusiastic about the SPIF process, were comfortable with the forms and were using them. However, we realized that we needed to do more than merely satisfy the customer in order to be successful. The health care industry is very competitive, particularly in the Minneapolis/St. Paul area, and we needed to make sure that our customers were not just satisfied, but delighted.

The next steps were not as easy as we thought they would be. We had a vision of what we wanted to do but we had taken too big of a leap for some of the QLT members. We needed to regroup and determine what the obstacles were.

Some initial obstacles

One barrier was the strong technical bias in the laboratory. We are very analytical by training. We focus on numbers, on output, on things that are easily measured.

The laboratory staff needed to move from producer-centered thinking to a Customer Centered Focus. Producer-centered thinking focuses first on the process, then the product (test results), and finally the outcome. We are very good at what we do, but we needed to "get out of the box" and realize that correct, timely and accurate results are not enough.

Education was essential

We started by educating the QLT members so that they could start looking from the "Outside In" or with customer centered thinking. A customer looks at the outcome first, then they look at the product, but they usually don't care about the process that was used to get the product.

The three rings of service

We used the "Three rings of Service" concept that is found in Chapter 2 "What is Customer Service" of Jim Clemmer's book, *Firing on all Cylinders* (See Figure 10). According to Clemmer, there are three rings of service. The inner ring is the basic product or service, the next ring is support, and the last ring is enhanced service, or the "delighters." At an inservice for the QLT members, we listed all the products or services that we do for our customers. We then placed these products or services in the ring we felt was appropriate. To our surprise, we discovered that most of our products/service are in the first two rings. We had only one item in the third ring. This was the "ah ha" that the QLT members needed. They began to realize that we do a good job and have a good product, but we need to do a better job on enhancing service.

Figure 10. The Three Rings of Service

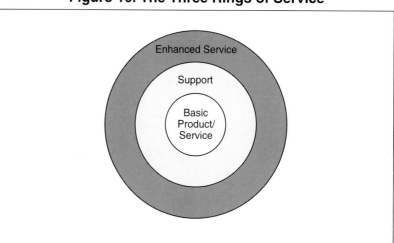

Identifying customers

Our progress has been slow but steady. From our many customers, the QLT members identified four groups of critical customers. These customers are:

- Coworkers
- Patients
- Providers
- Nurses

We needed to find a way to determine what our customers want. Customer surveys seemed to be the best measurement tools to determine the customer's needs and expectations.

Developing surveys

The QLT members determined "moments of truth," or key interactions with our critical customers, and formed subcommittees, one for each customer group. The customer teams then developed surveys, using the identified "key interactions" for each group of critical customer.

Staff participation

In addition to determining our customers wants, the QLT members also wanted to get participation by as many staff as possible. In order to accomplish this, a QLT member was assigned as a resource person to a site or department. The QLT resource person solicited a champion from each site or department and trained them in the use of each of the surveys (train the trainer). The QLT resource person also helped with any questions during the surveys and data collection.

The champions oversee the surveys at their site/department, the training of all staff and the collation of the survey data. All staff participate in the either the survey or in the collation of the survey data.

The site champions share the survey data at site/department meetings. If a gap in service at the local level is found, an Action Item form is generated and improvements initiated at the local level. The data is also sent to the QLT customer subcommittee that developed the survey, where it is combined with the data from all the sites/departments.

The customer subcommittee reports the results of the survey to the QLT. A SPIF is generated for any system-wide need or gap in service.

Current status

As of July, 1996, two surveys have been conducted. They are the coworker survey and nurse survey. Staff participation has been very good with several Action Items in progress.

The entire process will probably take the rest of 1996. The timeline for the surveys is to train the site champion, conduct a specific survey one month, and collate the data the next month.

Results from all the surveys will be published in the QLT newsletter. Survey results from the nurse, physician and patient surveys will also be published in the

Current status, continued

laboratory newsletter, "The Lab Oratory," with distribution to all HealthSystem Minnesota employees.

Some positive reaction

We are really starting to see some changes from all the staff. There is much more buy-in to fixing systems and less blaming of people. We asked our staff to tell us one good thing that they learned from a SPIF training session. When we read those comments, a small sampling of which appear in Figure 8, we realized that we're on the right track.

Figure 8. Employee Comments

"This system explains why this is a better place to work than my previous place of employment, it's worked even in its beginning and should improve."

"A good feeling about our job, an eagerness to be a good employee as there is reward for doing a job well."

"It's nice to know that if I take the time to think through a problem and fill out the form, something will be done about it; someone cares. I appreciate that."

Author information

Christine Kelly founded the Laboratory Quality Leadership Team in July, 1993, and served as the first team leader. She was the Director of Laboratory Operations for Park Nicollet Medical Center from 1992-1995, and is currently the Director of Laboratory for parent company HealthSystem Minnesota. Her current responsibilities include leading and managing all laboratory operations in 17 clinics and one hospital campus.

Elizabeth Lentz is currently a Regional Laboratory Manager at Park Nicollet Clinic HealthSystem Minnesota. She is responsible for the overall direction and supervision of eight clinic laboratory sites which include 34 employees. As the leader of the Quality Leadership Team for the past 1 1/2 years, Ms. Lentz has been instrumental in teaching Quality Improvement principles to staff and promoting a customer-centered laboratory culture.

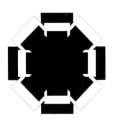

Putting the Patient at the Core of a Health Care Organization

Authors

Tom Tibbitts, President, Trinity Health Systems, Inc., Fort Dodge, Iowa

Sue Thompson, Vice President, Patient Support Services

Introduction

In January 1992, the Board of Directors and Senior Management Staff of Trinity Regional Hospital announced the beginning of a major new project called Patient Focused Care. This was a restructuring and Reengineering project with a purpose to: (1) examine the ways in which service was provided to patients, and (2) make recommendations on improving the organizational and functional relationships between departments and services within the organization, in order to provide patient care that is ultimately viewed as seamless by the patient.

Trinity Hospital: An overview

The hospital itself consists of a 200 bed campus with 75 active staff positions, and 1200 employees within the entire corporate system. It is a secondary level hospital, so things like open heart surgery or neurosurgery are not performed. This facility serves a very rural part of the country, with a population base of about 250,000 people in the service area. There is no other hospital in the area, so we have something of a monopoly in Fort Dodge.

Reasons for reengineering

At the time we first considered reorganizing, it seemed that just about every business and industry, including the health care sector, was looking to reengineer in some way. Even though there was not a lot of experience with reengineering in the industry, we made a decision to get out in front of that curve because we believed it offered great possibilities. It turns out we were right.

Patient Focused Care: Establishing a new organization

We began the process of instituting Patient Focused Care (see Figure 1) by establishing a Steering Committee (consisting of administrative staff) and a Design Team (middle management staff from all disciplines). Our organization also began to work with an Atlanta consulting group, the Patient Focused Care Association, in completing an Applicability Assessment. The purpose of this assessment was to determine the appropriateness of applying all the concepts of Patient Focused Care to our

Patient Focused Care: Establishing a new organization, continued

Figure 1. Patient Focused Care - A Calendar of Progress

❖ **January, 1992**
  ▶ Commitment to organizational restructuring by Board of Directors and Administrative Staff

❖ **March, 1992**
  ▶ Steering Committee and Design Team selected

❖ **May, 1992**
  ▶ Applicability Assessment completed

❖ **August, 1992**
  ▶ Initial Vision document as prepared by PFCA reviewed by Design Team

❖ **December, 1992**
  ▶ Design Team presented recommendations for restructuring organization to Steering Committee

❖ **January, 1993**
  ▶ Restructuring of organization approved by Steering Committee
  ▶ Care Center Administrators selected
  ▶ Nursing Governance Model approved and implemented
  ▶ Industrial Engineer joined TRH

❖ **February, 1993**
  ▶ Implementation Teams for each Care Center selected; weekly meetings initiated to complete detailed Care Center design

❖ **March, 1993**
  ▶ Quality Systems restructured to meet needs of Care Center environment

❖ **April, 1993**
  ▶ Care Center budgets for FY-1994 prepared by Care Center Teams and Administrators
  ▶ Implementation Teams determined appropriate care/service modalities as candidates for redeployment

❖ **May 1993**
  ▶ Plan for redeployment of services finalized by Implementation Teams for Care Centers

❖ **June, 1993**
  ▶ Business plans for each Care Center finalized by Care Center Administrator

hospital (Figure 2). In addition, the consulting group assisted us in *visioning* and *operationalizing* those concepts, resulting in dramatic differences in how we are organized as caregivers, how relationships between departments have changed, and how we have improved patient care processes hospital-wide.

Figure 2. The Applicability Assessment: Determining the Appropriateness of Five Concepts of Patient Focused Care

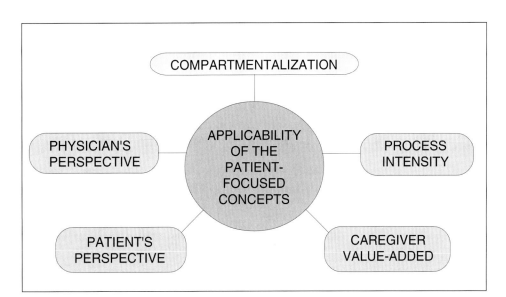

Applicability Assessment

The Applicability Assessment consisted of studying five components that would be influenced by the Patient Focused Care concepts (see Figure 2). Compartmentalization refers to how much structure existed that kept people from communicating with each other. Process Intensity refers to our analysis of various processes and our potential to integrate or streamline them. Caregiver Value-Added refers to how much time our caregivers were actually spending on direct patient care, as opposed to doing paperwork, for example. Patient's Perspective refers to a very important component: How many faces a patient would see, how long an average visit would last, and their overall satisfaction with the care provided. Lastly, the Physician's Perspective refers specifically to the ability of a physician to get their work done while dealing with pharmacists, radiology labs and nursing staff.

Results of the Applicability Assessment

The process of restructuring and reengineering the organization began when the Applicability Assessment was completed. The findings of this study were as follows:

- Communications between departments and caregivers revealed a highly compartmentalized organization.
- The simplest, most routine patient care processes contained multiple steps, but few steps actually related to the medical, technical or clinical care of the patient. The remaining steps were related to scheduling, coordinating and transporting the patient to finish the process.
- Less than half of the caregiver's time was being spent in direct patient care. A significant amount of time was being spent in documentation, scheduling, coordinat-

Figure 3. Findings from the Applicability Assessment

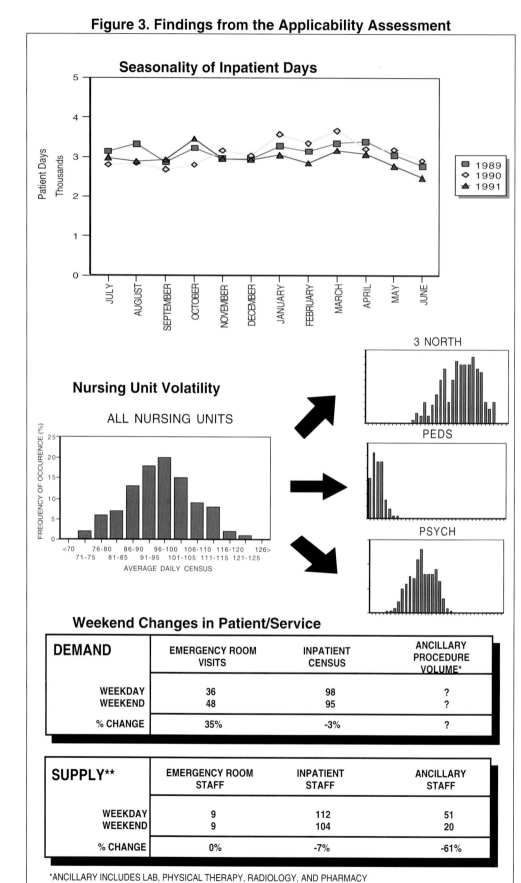

*ANCILLARY INCLUDES LAB, PHYSICAL THERAPY, RADIOLOGY, AND PHARMACY
**STAFFING DOES NOT INCLUDE MANAGEMENT OR CLERICAL RESPONSIBILITIES

Results of the Applicability Assessment

ing and completing system-imposed tasks associated with accessing supplies and care for patients not provided within the department.

• Continuity of caregiver assignment was not given priority in daily operations. In a study sample, a patient with an average length of stay of three days saw 48 caregivers.

• Physicians perceived many opportunities to improve turnaround times for Laboratory, Radiology, and Pharmacy services, and most commonly, were concerned about the lack of availability of a caregiver who was familiar with their patient when the physician was making rounds.

Figure 3, on the previous page, shows some of the assessment findings in a graphic format.

More findings from the Applicability Assessment

Highlights of the findings from the Applicability Assessment in Figure 3 are:

• In studying the seasonality of inpatient days, we observed very flat census lines when aggregating the hospital-wide census over the course of three years.

• Within the nursing units, we noted extreme volatility and recognized the challenge placed before each and every caregiver unit in managing staffing and census volatility.

• When studying weekend changes in patient service and the subsequent matching of supply with demand, we found many opportunities. Most significantly we noted the drop in supply of ancillary staffing by Laboratory, Physical Therapy, Radiology and Pharmacy on weekends during the same time that Emergency Room (E.R.) visits increased 35 percent and inpatient census was relatively stable.

Creating a Vision, aligning services

After analyzing the Applicability Assessment, the Design Team created a Vision for improving the provision of patient care. We wanted to align departments and services organizationally with caregivers and patients in need of the respective service. As each patient type was studied, five "operating levers" were determined: schedulability, predictability, length of stay, nursing care needs, and ancillary service consumption (Figure 4). Therefore, the Design Team recommended we align services by consumption patterns with patient types, in order to improve the organizational and functional relations within the hospital.

Creating the three Care Centers

As the process of creating Care Centers began to evolve, Care Center Administrators were selected. The Steering Committee established organizational goals and indicators in the areas of economics, service, quality, and environment. Three Care Centers (see Figure 5) were created based on an analysis of patient types cared for at Trinity and their common "operating levers." They are:

• **Care Center 1: The Family Care Center**

Patient type: Obstetrics, pediatrics, psychiatric.

Creating the three Care Centers, continued

Operating levers: Low schedulability, low predictability, low ancillary consumption.

• **Care Center 2: The Surgery Center**

Patient type: Surgery, recovery room, anesthesia, radiology, laboratory, physical and occupational therapy.

Operating levers: High schedulability, high predictability, high consumption of routine ancillary services.

• **Care Center 3: The Medicine Center**

Patient type: Emergency department, inpatient medicine care, pharmacy.

Operating levers: Low schedulability, moderate predictability, high consumption of routine and non-routine ancillary services, high need for information turnaround.

Figure 4. The Patient Focused Care Design Process

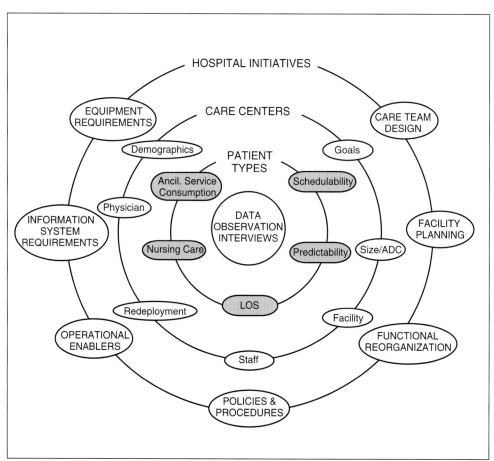

Figure 5. Two Organizational Charts: Before and After Reengineering

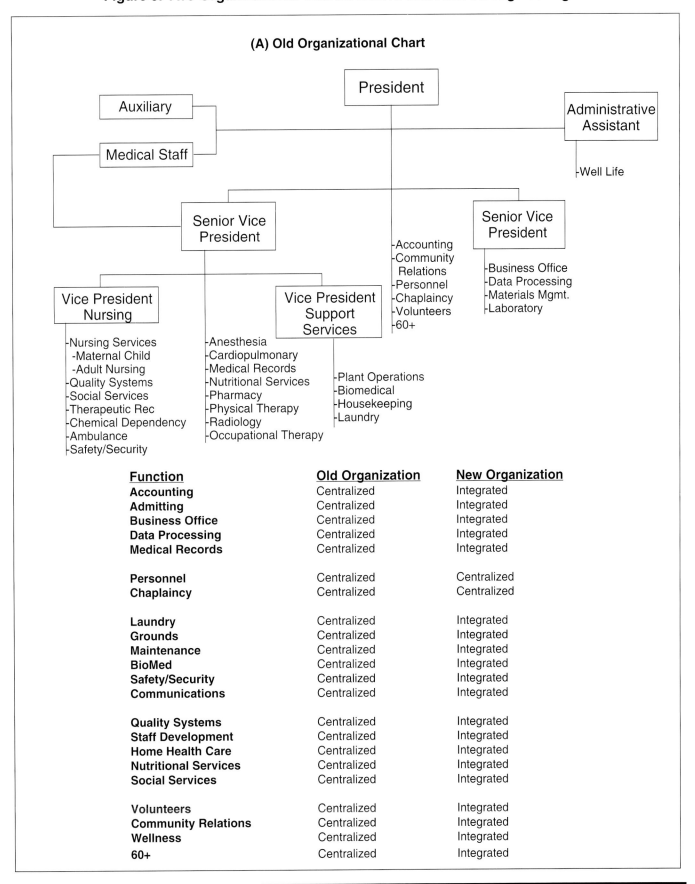

(A) Old Organizational Chart

Function	Old Organization	New Organization
Accounting	Centralized	Integrated
Admitting	Centralized	Integrated
Business Office	Centralized	Integrated
Data Processing	Centralized	Integrated
Medical Records	Centralized	Integrated
Personnel	Centralized	Centralized
Chaplaincy	Centralized	Centralized
Laundry	Centralized	Integrated
Grounds	Centralized	Integrated
Maintenance	Centralized	Integrated
BioMed	Centralized	Integrated
Safety/Security	Centralized	Integrated
Communications	Centralized	Integrated
Quality Systems	Centralized	Integrated
Staff Development	Centralized	Integrated
Home Health Care	Centralized	Integrated
Nutritional Services	Centralized	Integrated
Social Services	Centralized	Integrated
Volunteers	Centralized	Integrated
Community Relations	Centralized	Integrated
Wellness	Centralized	Integrated
60+	Centralized	Integrated

Figure 5. Two Organizational Charts: Before and After Reengineering (continued)

(A) New/Current Organizational Chart

PRESIDENT (T. Tibbitts)

CFO (B. Dixon)
- Business Office
- Data Processing
- Materials Mgmt
- Physicians Billing Service

Care Center #1: VP Patient Care Services (L. Hickey)
- OB/Nursery
- Pediatrics
- Adult Psych
- PHP
- Adol Psych
- Chem Dependency
- Rec Therapy
- Social Services

Care Center #2: VP Patient Care Services/CNE (M. Corkrean)
- OR/Rec. Room
- OP Surgery
- 3 North
- Skilled Nursing
- Anesthesia
- CS
- PT
- OT
- Laboratory
- Radiology

Care Center #3: VP Patient Care Services (F. Lineer)
- Emergency Serv.
- ICU
- 2 North
- Cardiac Rehab
- Renal Services
- Pharmacy
- Cardiopulmonary

- Accounting
- Community Relations
- Chaplaincy
- Volunteers
- Buildings & Grounds

- Employee Health
- Personnel
- Safety/ Security

- Home Health
- Corp. Health/ WC 2000
- 60+
- Well Life

VP Support Services (R. Kuhlman)
- Housekeeping
- Laundry
- Communications
- Nut. Services
- BioMedical

VP Pt. Support Services (S. Thompson)
- Quality Systems
- Staff Develop.
- Medical Records

Creating the multidisciplinary care teams

The Steering Committee also recommended a care delivery system that utilizes a multidisciplinary care team. The goals of a multidisciplinary care team include:

- Reducing compartmentalization, thereby improving the communication and cooperation between traditionally specialized caregivers.
- Improving the continuity of caregiver assignment to each patient.
- Improving the utilization of all staff.
- Reducing the lengths of patients' stays and general costs.

Applying a model for the nursing units

With the decentralization of Nursing in our Care Center environment, the Chief Nurse Executive assumed an even more important and significant role. The Professional Nursing Governance Model was established to assure that nursing care and practice would remain consistent across all areas within the organization. The model included a committee that addresses Nursing Education, Nursing Standards, Nursing Quality and External Regulatory issues. The Chief Nurse Executive is accountable for the administration of the Nursing Governance Model. Since professional services are decentralized into the Care Centers and Care Teams, a similar model for governance has been established.

Reengineering: an overview of the ongoing process

While the process of organizational restructuring and reengineering is an evolving one, it is clear that traditional departmental walls are being taken down and

Reengineering: an overview of the ongoing process, continued

new lines of communication are being established between ancillary service and nursing personnel. As a result, there have been significant changes and improvements in the processes of patient care.

We realized that although ancillary services are aligned with a respective Care Center (i.e., Laboratory with Surgery Care Center) the services may be consumed by patients from another Care Center. Therefore, we believe that if processes can be improved where the majority of the service is consumed, the benefits will rollover into the other two Care Centers. Regularly scheduled Care Center meetings include that particular center's administrator, their respective department managers, and representatives of all the ancillary services, in order to improve and clarify functional relations. This cross representation at the Care Center meetings is an important component in our efforts to improve the provision of care at Trinity.

As each Care Center meets to identify opportunities for improving patient care processes, some method is needed for determining the prioritization of and allocation of hospital resources to these activities. The Steering Committee agreed on criteria that would serve organization-wide goals. In the event of conflict between Care Centers regarding resource requests, the Patient Focused Care Steering Committee serves as the forum for resolving those issues.

Ultimate benefits of restructuring and reengineering

The process of restructuring and reengineering our hospital through Patient Focused Care has improved Trinity's ability to provide a more efficient, patient-centered type of care. A pharmacist has been moved out to each Care Center, supporting the work of physicians and nurses, and attending to patients' needs. Staffing of the emergency room has been adjusted to meet our patients' demand for care, which is particularly high on weekends. Continuity of care has been improved dramatically, especially in Care Center 3, so that the number of caregivers per patient has been greatly reduced. The patients are no longer seeing a dizzying number of faces when they come in. Preadmission testing and registration has been decentralized, so that patients are served by one or two caregivers, instead of five or six, prior to surgery. Departments have been consolidated and responsibilities have been combined. Each Care Center has a Quality leader, to put those ideas at the forefront of everyone's minds. A case manager model has been implemented, and we have seen better quality and a stronger team concept.

Evaluation and some results

There have been some significant, positive results in patient satisfaction, physician satisfaction and saving money at the same time. Evaluation of this process for providing care is conducted on an ongoing basis by the Steering Committee, with quarterly updates given to the Board of Directors to assess the appropriateness of this strategy. Relative to before the restructuring began, patient satisfaction has not changed significantly. There have been high marks in some areas, but since so much has changed,

Evaluation and some results, continued

and the process is still ongoing, it is difficult to assess their responses. Physician satisfaction has improved dramatically, since their access to information, services and assistance is much better. Financially, we are seeing some savings, but more importantly, there has been a real cultural change that will benefit everyone in the future.

Lessons learned

We have learned many lessons during this process.

• Resistance to change is alive and well. The solution to that is communication. We have tried to communicate with our employees as much as possible so that they know what we are doing and where we are headed.

• You must have everyone involved, and you must make things relevant to your staff. For example, our physicians did not get fully involved until they realized that the changes would affect how they dealt with patients and the rest of the staff. Then they decided to make it work.

• Teams are powerful. You need to listen to what your employees have to say.

• Leadership skills are critical. They are also everywhere; in various meetings and situations, different people emerged as leaders. That has been one of the great rewards of this process.

• Executive buy-in is essential. You must have the CEO directly involved with every aspect of a reengineering effort for it to work.

Author information

Tom Tibbitts and Susan Thompson are both from Trinity Regional Medical Center in Fort Dodge, Iowa.

Tom Tibbitts is president and CEO of Trinity Health Systems, within which there are seven subsidiary corporations including Trinity Regional Hospital. His positions at Trinity Regional have included Associate Director, Executive Vice President, and President. He's very active in his community serving on many boards of directors in a variety of professional and community organizations and activities.

Susan Thompson coordinates all the hospital-wide activities that are related to patient focus care at Trinity Regional Hospital. She's also responsible for administration direction to the Quality Systems and Staff Development Departments. She's involved in several associations whose membership include Quality Assurance and Nurse Professionals. Susan received her B.S.N. degree from Bishop Clarkson College of Nursing in 1991 and became a certified professional in Health Care Quality.

From Incremental to Breakthrough Performance

Author

Ellen J. Gaucher, Senior Associate Director & COO, the University of Michigan Hospitals, Ann Arbor, Michigan

Why Breakthrough Performance is Essential

The health care industry demands breakthrough performance

These are tough times for the health care industry. We always need to be thinking of more effective ways to compete. We should be taking care of what really matters—our customers and their satisfaction. We must be prepared for the future. We must continue to focus on world-class quality. Finally, and most importantly, we have to keep our energy level up, sustain our efforts, and keep our eye on the ball. The only way we can accomplish all these things is to move beyond incremental improvements to real breakthrough performance (see Figure 1).

Figure 1. The Road to World Class Quality

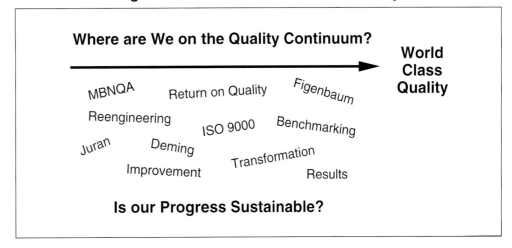

A transformational agenda is not the sole solution

Most organizations have something they call a *transformational agenda,* which focuses on overhead reduction, downsizing, employee empowerment, process redesign, and portfolio rationalization. I understand that these are critical issues, but I do not believe that this agenda is enough to reshape an entire organization and prepare it for the future.

Awards do not imply perfection

We've been successful at the University of Michigan Hospitals, and we've been recognized for that success. We won the State of Michigan Quality Leadership Award in 1994. We won the Health Care Forum's Commitment to Quality Award. We've come a long way in changing the organization, and yet I still see a lack of attention to the details we know have to be in place.

Leadership is essential to the continuing effort

I once believed that maybe in five years there would be an endpoint to transforming our organizations (see Figure 2). But transformation is regeneration—it's not a one time event. If somebody is not paying attention, always energizing and making sure that the passion is still there, then I think an institution will go into a denial phase. Leadership at all levels in the organization is critical if you're really going to have a transformation.

Figure 2. Organizational Transformation

Key components of organizational transformation:

- Continual regeneration
- Not a one time event
- Multiple strategies to reshape organization, re-skill people, and meet customer needs.

A new global field requires a three-part plan

When I became COO in 1987, I was looking for a way to wake people up and say, "There's so much more we can do, there's much more we have to do to be successful." Tichy and Devanna[1] said that because of triggers in the external environment, this new global playing field, there is now a need for a *three-act play* (Figure 3). On the left side of the figure are the organizational dynamics associated with each one of the three acts. On the right side are those individual kinds of things that need to happen in an organization.

Recognizing the need for change and the challenges ahead

Act One is the easy one, or so it seemed at the time: Recognizing the need for revitalization, transformation, and change; but also recognizing that there might be some resistance and an unwillingness to let go of some things that were done in the past; and all the while trying to avoid the quick fix.

Creating a new vision

Act Two: Create a motivating vision. Develop that vision and mobilize commitment. We began to make visible, sustainable change in my organization. About 20,000 people work on our campus, and many of them say, "You can't change a place like this."

        1. Tichy, Devanna, *The Transformational Leader*, J. Wiley & Sons, New York, 1990.

Figure 3. Organizational Transformation: A Three-Act Play

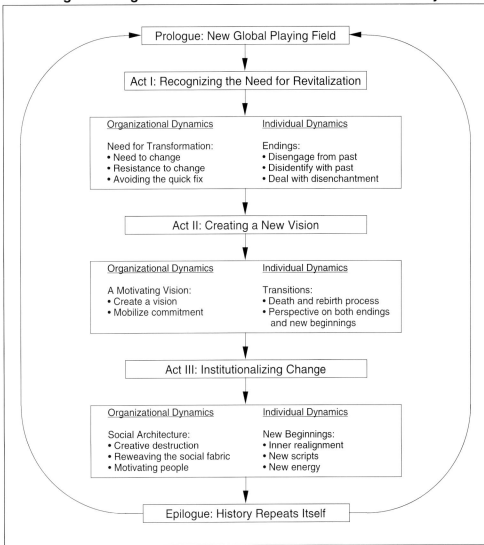

Creating a new vision,
continued

They're wrong. Turning around a crisis in a large organization is like trying to turn a battleship. It takes a lot of space and it doesn't happen instantly, and yet we are still making a change.

Institutionalizing change

Act Three: Institutionalizing change. This is where we began to lie to ourselves, saying things like, "We no longer need all the trappings of TQM. We have really imbedded the changes into the system, so we'll do away with the Quality Councils and we'll make sure that the management team always has Quality on its agenda. We no longer need the measures, consistency and attention that we've had. We can now move on to other things." The result was that the play stopped! Nobody seemed to realize that without that constant attention to detail, and following up on things, we did not reach a level of maturity—it was a level of nonexistence. I would walk around the organization

Institutionalizing change, continued

and hear people saying, "Is our Total Quality stuff over? Have we reached the end of TQM? Now we're doing reengineering?" That confusion and inconsistency concerned me. The play needs to repeat itself over and over again, every day, if it's going to help us really transform the organization.

Traditional methods of change do not prepare us for the future

We've restructured so many times it's pitiful. We've downsized, and hopefully we've done it in a humane and effective way. We've reduced the number of administrative people, but we're not small enough (see Figure 4).

We've also become better. We've reengineered processes. We've looked at reorganizing and reinventing our organization in many different ways. We've used the continuous improvement process to help us move ahead. But again, it's not enough.

Today we need to think about what health care should look like in the next 10 years. We need to think about how we are going to be different in the future, what that future is going to look like, how we are going to keep our attention focused on that difference.

Figure 4. Three Different Strategies for Transforming Organizations

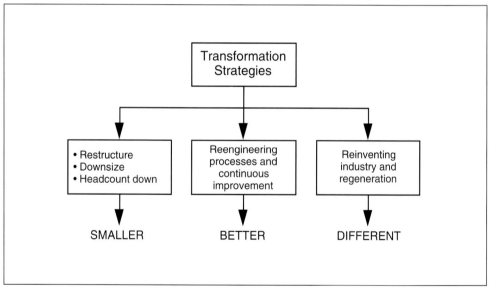

First order change

We seem to be very good at first order change (see Figure 5), which are incremental changes within already accepted frameworks. They are logical, rational, minor improvements, but they are reversible and do not change the system's core processes.

Figure 5. Three Levels of Change and Performance

First Order Change:
- Incremental changes within already accepted frameworks
- Minor improvements
- No change in system's core processes
- Reversible change
- Logical and rational.

Second Order Change:
- Change in systems
- Breakthroughs of large magnitude
- Revolutionary "leap frog" jumps
- Irreversible change
- Seemingly irrational change based on "out of the box thinking"
- New paradigm.

Breakthrough:
- All managerial activity is directed at either *breakthrough* or *control*
- *Breakthrough* = Change, a dynamic decisive movement to a new, higher level of performance
- *Control* = Staying on course, adherence to standard, prevention of change.

Example of first order change: Reduced waiting in Admissions

One of the first changes we made, and one that I think every hospital in the US has done, is reduce time in the Admissions waiting lounge. However, there are so many cycle time improvements that we can make in health care—waiting to see your doctor, waiting for your lab tests, waiting for your x-rays—that we could spend the next 50 years working on cycle time reductions and probably still not meet everyone's needs. We've not had any major changes in the system's core processes (or key processes). That process in patient care, doctor, patient, nurse, really hasn't changed too much at all. Reimbursement has changed, which is forcing us to operate in new ways, but we're still not looking at the real heart of the business and the direction that we need to move in.

Second order change

What we really need is second order change. These are changes in systems, breakthroughs and revolutionary jumps. They are irreversible, seemingly irrational changes based on "out of the box" thinking, and they create a new paradigm.

Example: Redefining Admissions

For example, having people wait in an Admissions lounge is an idea that has no place in our future. Why should we only reduce the amount of waiting time, when the whole concept of waiting should be eliminated? We now register 99% of the people by phone, and if they need urgent care, we bring the Admissions clerk to the hospital room and do it in the room. This was the first quality improvement project that we did back in 1987, when our patients were waiting an average of 180 minutes in the Admissions

Example: Redefining Admissions, continued

lounge. We now meet the people at the front door, just like at a fine hotel. You pay a lot more to come to your local hospital than you pay at a fine hotel, so why shouldn't we provide the same kind of services?

We need to achieve breakthrough improvements as well as incremental ones

We're still not seeing the breakthrough improvements that we need in our organization. We really need to focus on improvement of quality and improvement of systems, and make irreversible changes. We have to think about incremental improvement, but we also cannot be loading up the processes. The problem is that, particularly in health care organizations, people tend to crawl into their box on the organizational chart and become *square*. They stay within those boundaries, rely on all the controls in the system, and are not innovative or creative. We need to get people out of those boxes to really create effective changes. We need to break out of that square mind-set and move on to a more effective way of thinking. I believe that we have a long way to go in my organization and in the health care industry to create new paradigms.

Barriers to Breakthrough Performance

Getting to Breakthrough Improvement

I believe the reason that we don't achieve breakthrough improvement is because there are so many barriers, and the world is so chaotic for many of us. We have identified seven major barriers to breakthrough performance in our organization:

1. No perceived need to change.
2. Lack of systems thinking.
3. Avoidance of the need for planning.
4. Unclear, undirected, unmeaningful and untracked goals.
5. Unsupportive culture for change.
6. A weak customer focus.
7. Ineffective training.

Barrier 1: No perceived need to change

Change creates fear that must be channeled into a sense of urgency. We all like to say that we are change agents, that we make things happen in our organizations. However, there have been times when I've thought to myself, "We're going to do *that*? *That* doesn't sound like it's going to work too well." If I'm thinking this way, what must other people in the organization be feeling? It might be understated, but there's a tremendous amount of fear in an organization when it comes to change. When we sense that there's no perceived need to change, most of us jump up on the soapbox and give a rousing speech, which might make everyone feel good for a little while, but it won't help the organization change. I think we need to focus on creating a sense of urgency about doing it in little steps where people can follow what we'd like to do, and where we can help people make a difference.

Barrier 1, continued

The Trap of Success

There also is a problem with what I call the *trap of success* (Figure 6). We've been successful at Michigan for many years. We've made good margins, we've had a lot of money in the bank that we're saving for a rainy day, and we're feeling pretty good. But when you have unparalleled success, you also start to develop symptoms. You tend to become complacent, be very content with current performance, have an internal focus, low creativity and innovation, and allow bureaucracy to creep up everywhere. These symptoms can lead to a major disease. You could become the highest cost hospital in the state. You could become an institution that doesn't focus on its customers, which could be the executives, the physicians, or the corporate leaders in the organization. You could have no innovation and creativity. This would be fine if there were no external challenges, but in our changing world it is easy to get caught up in a *death loop*. You'll see a decline in performance, you'll continue to do the same thing, things get worse and worse, and you're caught in a downward spiral. You cannot keep doing the same things and still expect to see change.

Figure 6. The Trap of Success

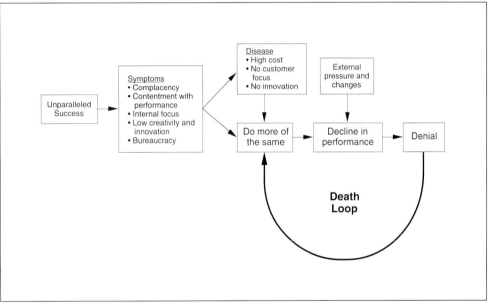

Create a Sense of Urgency

So how do you create that sense of urgency, to get people to do things differently? One way is to focus on urgent, short-term goals to get people energized and focused in a new way (see Figure 7). Instead of talking about your problems and shortcomings, and fighting over whether the data is valid or not, think about creating that sense of urgency. Short term results are all so important. They give you a chance to celebrate, feel good about things, reach a plateau and then be able to go on to the next

Barrier 1, continued

level. It's easier to jump-start an organization that's moving in the right direction, rather than one that's languishing, where you're not able to get people moving at all.

Figure 7. Creating a Sense of Urgency

Emphasize:	Instead of:
• Urgent goals	• Problems and shortcomings
• Short term results	• Long term or strategic focus
• Accountable managers with ownership for the process.	• Staff expert.

Give Managers Support and Authority

Another way to help people move in the right direction is to make sure that you have accountable managers. It is essential that you give them the authority, money, and tools to get the job done themselves, instead of just relying on experts. Consultants might come in and get things moving in the right direction, but when they go away and people start saying things like, "Is this program over? Aren't we still doing this other thing," it's up to the managers to keep them focused.

Barrier 2: Lack of systems thinking

Focus on improving as a system, not as individuals (Figure 8). A lot of people say they're systems-thinking, but if that is the case, then why do I hear everyone talking about personal or departmental goals? I was at a budget hearing where we had some big variances in the organization, and three of the executives, almost in unison, said, "My department is doing just fine, thank you very much." It's terrific if your little group is doing fine, but what about the rest of the pieces? Aren't we all in this together, and shouldn't we also be looking for solutions that go across the boundaries? We really have to be clear about what we want for the organization. We must prioritize our actions and the impact of those actions in decisions on the organization, not just on ourselves or on our divisions. Dr. Deming told us a long time ago that a system without an aim is not a system. When you have a group of unrelated departments and divisions doing their own thing, you're not going to make the same kind of progress that you can make when you're all working together. And unless we're doing well as a system, we're not doing well at all.

Figure 8. Barrier 2: A Lack of Systems Thinking

- People are primarily focused on personal or departmental issues
- There is a lack of "system" focus or aim
- Must prioritize organizational impact of actions and decisions.

Barrier 3: Avoidance of the need for planning

Planning and communication are critical. I have heard so many excuses about why we cannot plan (Figure 9). However, if we want to empower people, they have to know what they're supposed to be working on and where they're supposed to be going. Have you have ever walked around behind a team of examiners that's asking your employees: "What's the mission of the organization? What are your values all about? What particular goal are you working on that's going to help the organization do better with their key processes? What's a key process?" It can be downright frightening when you hear some of the answers. We think we've done such a good job, but we're not out there, up front as leaders, talking about the plans for the organization, the direction that we're moving in, or why it's so essential for everyone to know where we're going.

Figure 9. Barrier 3: Avoidance of Planning

Excuses:
- Don't have time; too busy
- Things are changing too fast
- Can't predict anyway, so why bother
- Whatever will be, will be
- Don't feel empowered
- Someone else will do it.

Some Key Leverage Points for Planning

We've tried to help people use leverage points to think and plan more effectively, to help their departments, divisions, managers, and supervisors, and move ahead. In our organization, there are several things that we think are critical (see Figure 10):

- Key service factors. What are the things that customers are complaining about? We've spent millions of dollars on defining customers and their requirements, trying to fix things, and every time we have a focus group we learn something that we didn't know before. The closer that we get to our customers, the more we find out what they really need and require. So we ask people to make sure they understand the chief complaints in their areas, and to do something to help that's visible to the customers.
- Diagnosis Related Groups (DRGs). What are the key, high volume, high cost DRGs in your division, and what are you going to do about it? How are you going to work on those? How are you going to reduce the cost and improve the quality?
- Cost per unit of service. How can we reduce the cost per unit of service, which is critical today in health care. With more HMOs and cost cutting measures, we have to be able to compete on cost.
- Budget overruns. Where are you in variance, and what are you doing about getting it in line?

Barrier 3, continued

Figure 10. Questions for Leverage-Point Thinking

- What are the key service factors customers complain about?

- Which DRGs (Diagnosis Related Groups) are high volume, high cost, and most in need of attention?

- How can we reduce cost per unit of service?

- Where are the biggest budget overruns?

Anticipate, Innovate and Prepare for the Future

In the past it's been OK for us to take up where our elders left off, to choose a role model, study them and think about learning and managing in the same way. Today that's just not possible. The world is changing so fast, no matter what industry we happen to be part of, that just doing what someone else did isn't going to help you. You need to be thinking ahead. Wayne Gretzky was interviewed and asked why he is such a successful hockey player, why he is able to score so many goals. He said, "I skate to where the puck is." Now how can we teach our managers to be able to skate to where the puck is, to anticipate the changes, the needs of customers, the requirements of customers, the need to make the organization more successful?

Barrier 4: Unclear, undi-rected, unmeaningful and untracked goals

Create goals that are clear, meaningful and attainable. We have too many goals, many times they're vague, unmeasurable, or people don't understand them (Figure 11). We need to make sure that everyone understands how those goals interrelate in the organization. One day I was with the radiology department, and it finally struck me. I said, "One of the key processes we want to improve for the whole organization is diagnosis. How can we diagnose patients quicker and speed up the treatment process?" I asked them to think about processes in a triangle, all leading up to our goals as an organization, and I asked them what they could do to help. It was amazing. One of the chief radiologists said, "We can turn forms around, read them faster." One of the technicians said, "We can get people through the radiology suite in a reasonable time and cut down retakes." One of the clerical people saw that they could get the films back to the file quickly so that the physicians and nurses on the floor could use them for diagnosis. I took the triangle model to other department meetings, and it made sense to people.

Many times we talk about concepts. They're very complex, but we tend to think everybody knows what we're talking about. By the simple use of that triangle, showing how everything feeds up to that organization goal, I finally had the departments understanding how this could work in our organization.

Barrier 4, continued

Figure 11. Barrier 4: Undesirable Goals

Avoid goals that:
- Are too numerous
- Are too future oriented
- Are elusive, vague and unmeasurable
- Have unclear work plans for goal attainment
- Have poor methods of tracking progress
- Have unclear accountability.

Reward System Must be in Line With Goals

Another issue is that we say we have five or six key goals, but the only ones we really reward are those that are financial. I can tell you a story about a hospital I visited, and the people in the intensive care units had done a fabulous job of reducing the length of stay. The problem was that this was not in a very competitive health care market, and the payment basis was still by fee for service: The more you do for patients, and the longer they stay, the more money you get. Well, after they had taken five days off the length of stay on a particular complex case, the clinicians were sitting around the table as the CFO said, "Look at the budget. Look at the revenues. We're $3 million under. What are you people doing?" They were doing what they had been asked to do; reduce the cost per case. I sat there mystified, wondering why the CEO and the CFO didn't see it. The workers were thinking about the future. They were reducing costs now so that they could bid on HMO business or contracted business, and basically stay afloat.

Losing revenue for a few months while the organization is changing is fine, that's the mind-set you want your clinicians to have. How do you think those clinicians felt when they left the executive table? Confused, depressed, dismayed. The financial goals somehow had superseded innovation, change, and moving ahead. It's sad. This brings me to another sad issue, and that's the cultural barrier in an organization.

Barrier 5: Unsupportive culture for change

Control and resistance must be overcome. In many instances, our cultures resist change with passion. There are two things that I think come into play: Control, all of our desires and abilities to control, and then the resistance to change.

Managers are Under a Large Amount of Stress

It is a confusing environment for managers today (Figure 12). Study after study has shown that managers' activities are characterized by brevity, variety, and discontinuity. They're hearing their leaders talk about things that don't make sense. They're bombarded on a daily basis with whatever the change strategy of the month seems to be. They can't understand how these programs relate together. Therefore, they tend to check

Barrier 5, continued

out, hide out, hope that this is all going to blow over, that all the talk about budgets is going to go away and they won't be threatened any more.

I was watching a football game recently—the quarterback threw three incomplete passes in a row, and the team had to punt the ball away. But when he went to the sidelines, the coach put his hands on his shoulders, and he put his forehead on his helmet, and he encouraged him. How often do we do that with our managers? We need to give encouragement to people, help them get to the next plateau, and do it on a regular and recurring basis.

In times of turmoil, people tend to hide out even quicker. They don't want to attract any more negative attention. If this happens all across the organization, and you're seeing this kind of denial and resistance, you're going to go nowhere with your quality program. And when times are very turbulent, I think each manager is working harder and harder to be more and more controlling, so they can feel more comfortable. This is the wrong attitude—we should be encouraging people to step out, to reach out, to get to that higher playing field, to skate to where the puck is. We cannot have people retreating, retrenching, and hiding out; that contributes to the resistance to change.

Figure 12. Today's Environment is Confusing for Managers

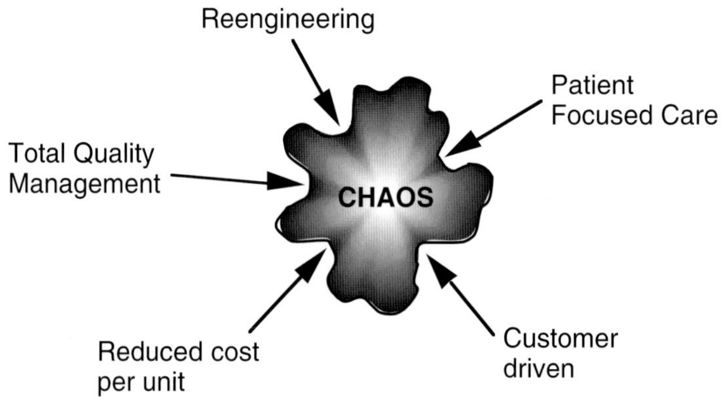

The rate of change creates confusion because:
- Many new programs don't relate to each other well
- Managers are overwhelmed with data.

In times of turmoil:
- The sense of chaos and unmanageability grows
- Things seem out of control
- Leads to anger, discouragement, and doubt about personal effectiveness.

When times are turbulent:
- Managers work harder to try to control and bring order to their lives.

Barrier 5, continued

You Must Have Action to Move Ahead

We have a tendency to be perplexed with this resistance, but it's actually a very human thing. We should expect it, and find out how we can encourage people to get involved, get them excited about changing their job, changing the organization, changing their processes. We shouldn't be just standing back and saying, "Oh, there's resistance to change." We should anticipate it and deal with it if we want to be successful. Juran said that we keep managers so busy, and keep them focused on that desire to control, that they have no time for breakthrough. We should be counseling, coaching, thinking about breakthrough.

We've done cultural audits and learned three things: (1) We've had a strong desire to be perfect, after all we are THE University of Michigan. (2) People who don't make mistakes don't do much else that's exciting, either. They tend to follow the normal routine so that they don't get punished. (3) We publicly chastised a lot of people for screwing up, and then we wondered why people weren't taking risks.

We are changing ever so slowly, but if you ask people behind closed doors, "What are the rewards here for stepping out and doing creative and innovative things?" I think you'll find that there's still some fear and some inability to move ahead.

Barrier 6: A weak customer focus

In the health care industry, one must always focus on the patient as a customer, even though understanding what the patients and the customers want is very difficult (Figure 13). We still really aren't as in touch with our customers as an industry as we ought to be. We have to back up from looking at our customer satisfaction surveys, and ask some real deep questions. I've been talking about health care as a non customer-friendly industry for a long time now. Many hospitals treat people like the state treats prisoners. They take away your clothes, they put you in a johnny or uniform for a day, or week, or month. They put you in a room with a stranger, they take away all your valuables, they tell you when you can eat, when you can take care of other bodily functions, and they think this is customer friendly. So what do hospitals do to make up for it? I hear people telling me things like, "These are the big changes we made in our organization. We now deliver the paper of choice to the customer in the morning." Most of them can't even read! They're hooked up and they're in for such a short period of time that that's not going to help one bit. We have to work on making people feel better about being in the hospital, about reducing the fear and the torment of people. We need to better define what people want from us. We have to put the customers in the primary position in the drivers' seats, instead of where they've been in the past, which is in the back seat.

Barrier 6: A weak customer focus, continued

Figure 13. Barrier 6: Weak Customer Focus

Developing a strong customer focus requires:
- Defining customers
- Defining requirements
- Assessing and closing gaps based on results
- Future orientation.

Barrier 7: Ineffective training

Don't just train for its own sake; train effectively and efficiently. I used to joke and say that in the early days of TQM we trained anything that would move. We took a couple of people from each department, trained them, and sent them back to their department. When they left the training program after five days, they were so excited, they had all the tools they really needed to change the organization. But when they got back to their jobs, their boss said, "Oh, you were at one of those Quality programs? Well don't think you're going to do that stuff here. I don't think it will work." It took us a long time to figure out that we needed to get into those departments and train all the people together, have them work on departmental problems, make them understand the concepts more effectively and appropriately about processes and the key processes, and show them how to identify gaps, set up teams around them and close them up. If you don't teach people to use the tools and techniques with something that's meaningful to them, their skills, knowledge and insights will deteriorate, and you will eventually have to retrain them.

Overcoming the barriers to move from incremental to breakthrough performance

In addition to identifying the barriers, there are several other things one must do to move from incremental performance to breakthrough performance:

1. Develop the role of leaders:
   - Create a sense of urgency, and answer the question "why change?"
   - Create, craft and communicate the vision.
   - Set stretch goals—use benchmarking and goal attainment to challenge staff; also use short-term focused goals.
   - Encourage pilot projects with PDCA cycle to stimulate small wins and constant actions. Success breeds success!
   - Focus the change process by pulling all the pieces together and actively supporting the learning process.
2. Create a change management "tool box." Include the following:
   - Problem solving skills
   - Seven management tools
   - Seven quality tools
   - Benchmarking process and skills
   - Reengineering
   - Simulation
   - Mathematical modeling
   - Systems.

Overcoming the barriers to move from incremental to breakthrough performance, continued

3. Use teams to manage. Include the following:

- Key process teams
- Self-directed work teams
- Benchmarking teams
- Reengineering/redesign teams
- Clinical assessment teams
- Management teams.

4. Focus the change process by pulling all the pieces together and actively supporting the learning process.

5. Assess the skills that are needed to change your organization, value innovation, creativity and flexibility, and recognize change is a constant concern among your employees.

6. Focus on results instead of activities. Many organizational change programs mistake means for ends and process for outcome.

Conclusion: It all starts with the individual

Achieving breakthrough is like any other major strategy. The process begins with the individual (Figure 14). You must begin to think about organizational values, goals and objectives that are above your own personal and departmental issues. You must commit your personal energy for change, and create the burning platform, whether it's at the department, the division, or the executive level. You must craft and communicate the vision of where you want to go with your organization, with your team. You must set stretch goals and expectations for people so they know what they're working towards, and they can celebrate minor successes along the way. You must personally focus on customers. And finally, it's up to you to keep that pressure on, and demonstrate continuous commitment to your actions and decisions. It's just like everything else in life: The transformational process begins with each one of us, and when we commit to making sure that we move from incremental performance to breakthrough, we can have a major impact on our organizations, and make a real difference.

Figure 14. Achieving Breakthrough Performance Begins with the Individual

- **You** must step above personal interest, consider what is best for the long term for your organization.

- **You** must commit personal energy to:
    - Create the sense of urgency
    - Craft and communicate the vision
    - Set stretch goals and expectations for change
    - Personally focus on customers
    - Keep the pressure on.

- **You** must demonstrate continuous commitment through actions and decisions.

Author information

Ellen J. Gaucher is Senior Associate Director and COO of the University of Michigan Hospitals, an 886-bed academic health center with several satellite facilities and over $550 million in annual revenues. She has more than 20 years of experience in senior management positions in health care organizations, and is a judge for the Malcolm Baldrige National Quality award. Since 1987, she has led the total quality process for the University of Michigan Medical Center, and most recently has been charged with primary care development for the hospitals. Gaucher is the coauthor of two books, Transforming Healthcare Organizations: How to Achieve and Sustain Organizational Excellence *(1990, with R. J. Coffey) and* Total Quality in Healthcare: From Theory to Practice *(1993, with R. J. Coffey). She has also written numerous articles and book chapters, and lectures internationally on hospital management, systems development and quality improvement.*

Hoshin Planning In Health Care: Current Initiatives At Two Medical Centers

A Case Study Discussion Paper

Authors

Geoffrey Crabtree, Vice President, Strategic Planning and Market Services, Methodist Healthcare System, San Antonio, Texas

Dona Hotopp, Director of Health Care Services, GOAL/QPC, Methuen, Massachusetts

Owen McNally, Director of Total Quality Management, Our Lady of Lourdes Medical Center, Camden, New Jersey

Editor's note: defining hoshin planning

Hoshin Kanri was developed in Japan in the mid 1960s and can be translated as policy planning and deployment, or management by policy (as opposed to management by objectives, for example). It is a planning system that points the organization in the right direction, with a strong focus on setting organizational targets together with the means to reach the targets. The hoshin planning system involves continuous improvement of planning. In its simplest form it involves a plan, execution, and audit. In its more detailed form it includes a long range plan (5-10 year vision), a one-year plan, deployment to departments, execution, monthly audits, and the CEO's annual audit.

Hoshin planning is recognized as having many advantages over traditional planning. It is thorough, data driven, and participative. Individuals make plans that are tied into a company vision, diagnose the processes, compare actual and target results, and analyze the cause of any gaps. The tools used are quite simple, easy to teach and use.

This paper is presented in a roundtable format

The inspiration for the style of this article comes from Dr. Akao's book, *Hoshin Kanri*. Dr. Akao promoted and taught Hoshin Kanri throughout Japan and the U. S. In the last chapter of his book he has a roundtable where he brings in Hoshin champions who talk about what's really going on and how they're really doing this. This paper, of course, is not a real roundtable but that's the sense we're trying to convey and the format that we're using.

The basis of this information was a highly praised roundtable presentation at GOAL/QPC's 11th Annual Conference in Boston, November 16, 1994.

Preface

Hoshin Planning was introduced in health care in 1990 and has been successfully used by health care organizations over the past five years. For two pioneering medical centers, Our Lady of Lourdes Medical Center in Camden, New Jersey and Methodist Healthcare System in San Antonio, Texas, Hoshin Planning has become "the way we do planning" and a means to address the rapidly changing health care environ-

Preface, continued

ment. Both organizations used the basic model of Hoshin Planning, starting with a Vision and 3-5 year plan, although with somewhat of a difference.

Our Lady of Lourdes started with Hoshin Planning for two reasons:

1. They had a Strategic Plan but wanted to make the Plan more customer-focused.

2. They were using the GOAL/QPC Total Quality Management model, which starts out with the customer in the center and then has Daily Management, Hoshin Planning, and Cross-Functional Management (see Figure 1).

Figure 1. GOAL/QPC's TQM Wheel

Our Lady of Lourdes demonstrates that one can "jump in wherever you are" and do "what works for you." They started with a Strategic Plan and then developed annual objectives. It was some two years into the process before the Vision, the true Vision for the organization, was developed. So it is possible initially to not have a clearly enunciated Vision. One can start the Hoshin planning process in order to focus on key strategies, objectives and goals.

Methodist Healthcare System, however, approached things differently. They used the Hoshin Planning process to develop their Customer-focused Strategic Plan.

In summary, one started with its strategic plan and one used the Hoshin process to develop the plan. Both organizations picked one annual objective to deploy each year. Both do an Annual Review, looking back and asking, "shall we keep the same objective or should we change our objectives?" (One of the key facets of Hoshin Planning is applying the (Shewhart/Deming) PDCA (Plan–Do–Check–Act) Cycle with regular reviews and Annual Reviews–actually using PDCA in the whole planning process.) Using the same model, but with different variations, both organizations started in similar time frames–1991/1992.

**First, an overview by
Dona Hotopp, GOAL/QPC**

The story of Hoshin Planning in Our Lady of Lourdes Medical Center and Methodist Healthcare System is very rich, both in the technical side and the people side. In this paper the authors will address the following items:

I. Hoshin Planning To Meet Current Health Care Challenges

• <u>Alignment: examples and results</u>

Hoshin Planning has brought the hospitals' strategic plans to life. It has enabled both medical centers to focus on key strategies to meet customer needs, and then to create the vertical alignment needed to mobilize divisions, departments and individual employees in developing the targets and means to implement the strategies. Examples include reducing length of stay, implementing total quality initiatives, and improving hospital/physician relations to prepare for managed care.

• <u>Cost Reduction: process and results</u>

Though cost reduction was not a strategic objective of either organization, cost reduction and revenue growth have been results of strategies adopted through the use of Hoshin Planning techniques. Cost reduction through improvement of health care delivery has been documented. How to use the Hoshin process to meet the continuing pressure for cost reduction will be discussed.

• <u>Empowerment: role of Hoshin Planning</u>

True empowerment is created through Hoshin Planning because employees know the strategic direction of the medical center and see how their work and their improvement efforts fit with customer needs and the medical center's strategy. Examples of focused empowerment, planning becoming a part of daily work, and other "soft side" benefits abound in both organizations.

• <u>Integration: how Hoshin Planning can help</u>

Methodist Healthcare System is using Hoshin Planning as a key integration methodology in a major new joint venture partnership with Columbia/HCA. Our Lady of Lourdes is using it to implement new strategic initiatives in developing their integrated delivery system. After three years of experience with Hoshin Planning, both organizations expect the process to be more focused and to help them in meeting the current integration demands of the industry.

II. Hoshin Review

• <u>Monthly and Annual Review Process</u>

The Hoshin review process is one that eludes many organizations in any industry. Our Lady of Lourdes used monthly review meetings and an annual "check (C)" in the PDCA cycle. Methodist annually reviewed the Hoshin objective as well as other facets of strategic planning. The benefits of the review process and plans to improve this process at both organizations will be discussed.

Overview, continued

III. The Big Picture

• Key Success Factors

The key success factors are unique to each medical center. The most important steps each organization has taken and what they would do differently if starting over today will be addressed.

• Major Challenges

The article will conclude with the major challenges ahead for improving Hoshin deployment and for using Hoshin Planning to react quickly and well to changing external forces.

Background on Our Lady of Lourdes Medical Center

Demographics: Our Lady of Lourdes is a 375 bed tertiary care hospital in Camden, New Jersey. It serves as the regional center for cardiology, dialysis and transplantation, prenatal care, and rehabilitation as well as maintaining family health and wellness centers. It is affiliated with the University of Medicine and Dentistry in New Jersey. The Medical Center is the 1995 recipient of the coveted Foster G. McGraw Award recognizing its achievement in community service, after being a finalist for the Award in 1992 and 1994. In addition it has received several awards in the last few years for innovation in access to the health care system within the community.

Hoshin Planning Purpose: Lourdes initiated Hoshin Planning to solve a major problem affecting customers and revenue: Lourdes was unable to meet the demand for cardiology and other emergency patients' needs to be admitted to the medical center. More capacity had to be provided without physically expanding the medical center. Hoshin was also seen as a method to activate and energize customer focus throughout the medical organization and to link operations more closely to the strategic plan. The organization has used the *Management and Planning Tools* in every phase of Hoshin Planning. These tools have become part of the fabric of their Total Quality Management implementation.

Hoshin Objective: *To expand effective capacity by reduction of length of stay through improved quality monitoring and physician utilization* was the Hoshin objective from 1991 through mid-1994. This objective was reviewed and progress tracked. In 1993 a numerical goal was added: reduce length of stay to 7.5 days by 12/3/93. Improving quality monitoring and physician utilization were the areas of activity. The results were reduction of length of stay and increased capacity. In 1994 the medical center moved that objective into its daily management and initiated a new objective involving the development of an integrated delivery system.

Getting started

The whole object was to integrate Total Quality Management into the way of doing business at Lourdes and use Hoshin Planning as part of the process. It was a learning process to apply this in health care. Lourdes started out with a retreat, reviewing the mission and Strategic Plan, looking at the Vision, looking at key Vision elements

Getting started, continued

to move them forward into the future, then defining which would be the key Vision elements they wanted to work on, and using the Management Planning Tools. They used Affinity Diagrams, Interrelationship Digraphs, and Prioritization Matrices to define each one of these steps.

Breakthrough objective

The key breakthrough area to work on was identified: *Clinical Quality Improvement*. This focus was further reduced to *Increasing Effective Capacity* (Figure 2). One of the issues that was troubling at the time was more demand for services than was possible to fulfill. People, especially cardiology patients coming through the Emergency Room, weren't able to get into the hospital. Increasing effective capacity would help them in their clinical quality, focusing on physician utilization and quality monitoring. Their breakthrough objective, developed after three days of work, was defined as *to expand effective capacity by reduction of length of stay through improved quality monitoring and physician utilization* .

Figure 2. Hoshin Generation Flow

Diagram reprinted from *Breakthrough Leadership*, © by 1995 American Hospital Publishing, Inc., page 154.

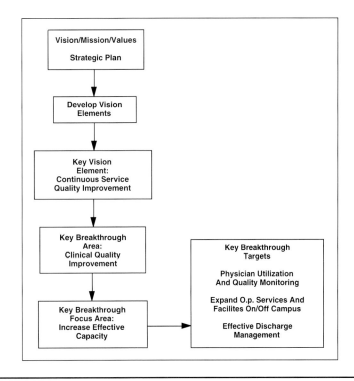

Initial teams

Two teams were formed, composed of vice presidents, department heads and clinical chiefs. The teams focused on "how are we really going to accomplish our objective?" One worked on collection and analysis of quality data, and the other on reduction of complications affecting length of stay. The VPs came together afterwards to see what the teams had generated. They had come up with 150 means to accomplish their objective. The VPs then took ownership of the key means that they felt were important to accomplishing the goal of *Increasing Effective Capacity by Reducing Length of Stay*.

Initial teams, continued

Each VP took ownership of a group of these means to accomplish the key objective and developed cross-functional teams to work on specific areas of concern. That was an interesting application.

Cross-functional teams

The graphic in Figure 3 shows Alignment. Each area or "bone" of the fishbone diagram represents a cross-functional team with a process owner indicated by a department in the box. These were the means that were being used by each of these teams to accomplish *Reducing Length of Stay Through Physician Utilization Quality Monitoring*. Everyone in the organization started to become aligned in accomplishing something that was important and visible, with everyone working toward the same purpose. That was what they thought was very, very significant. A numerical target was set in July based on capability and everything is measured month by month, to see how they're doing. All of this is made visible for everyone to see what they're accomplishing.

Figure 3. A Fishbone Diagram is Used to Picture the Organization's Alignment Toward its Hoshin

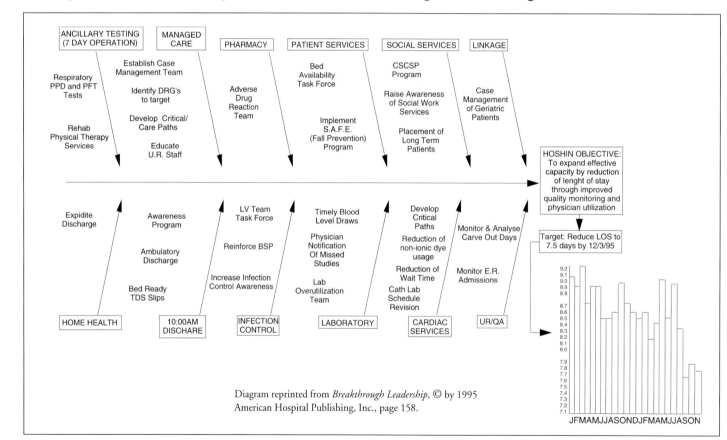

Diagram reprinted from *Breakthrough Leadership*, © by 1995 American Hospital Publishing, Inc., page 158.

Significant improvement is achieved in cardiology

The chart in Figure 4 shows some of the results. Through the leadership of the Chief of Cardiology, working on Critical Paths and Lab Utilization, they were able to treat more patients in fewer days.

Results in cardiology

Diagram reprinted from *Breakthrough Leadership*, © by 1995 American Hospital Publishing, Inc., page 162.

Figure 4. Cardiac Cases And Patient Days By Quarter

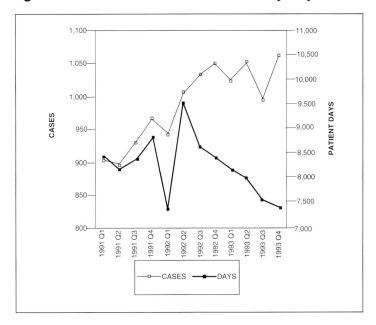

Process review

The chart in Figure 5 gives you the picture of how they went through the Hoshin Planning process over a period of a couple of years. I've described what happened across the top. They also used the *Catch-Ball Process* (a formal, vertical, communications), once the VPs had taken ownership of the means.

The Catch-Ball Process was used with Department Heads to decide what will be done in every department to achieve the goal. Deployment actually happened within a year. There were matrices that showed outcomes, responsibilities and time-lines. The whole plan was made visible. Everyone knew what they were working on. Quarterly meetings where held to review how they were doing, and then they achieved the results that are indicated here. One result was a bonus to all employees. I found it very impressive that the President of the hospital in the bonus letter to all the employees said, "we're sending this to you because you helped us accomplish our Hoshin objective." Everything was tied together. The reward was tied to the work. The work was tied to the means. The means was tied to tactics, and all going back to the Vision. In the coming year there will be new Hoshins generated which will involve everyone in the organization.

Background on Methodist Healthcare System

Demographics: The Methodist Healthcare System is now made up of five acute care hospitals and two ambulatory surgery centers. This newly created health care system is the result of the coequal partnership joint venture between the 573-bed Southwest Texas Methodist Hospital and the four Columbia/HCA hospitals, all located in San Antonio, Texas.

Hoshin Planning Process: Methodist Hospital, a few years prior to the joint venture, began using Hoshin Planning to address future potential financial considerations and to improve their strategic planning process. After initiation in the Deming

Figure 5. Hoshin Process Flow And Time-line

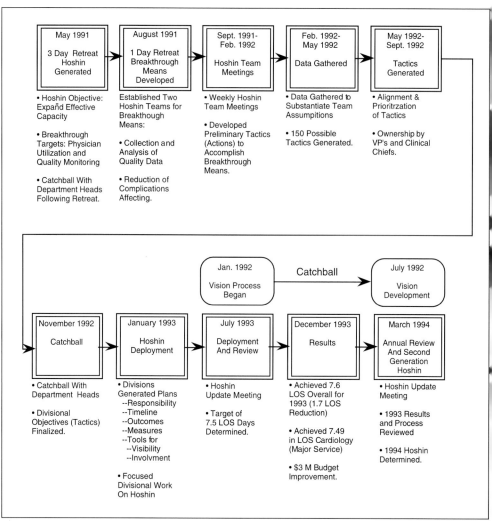

Background on Methodist Healthcare System, continued

philosophy, the leaders wanted to adopt a customer-focused approach to long-term planning that involved internal customers at all levels, as well as a specific focus on external customer demands and environmental trends. Hoshin Planning met their planning needs and provided the integration of goals, and a structured organizational communication plan, to improve their overall process. The "Management and Planning Tools" have been an integral part of the Hoshin Planning process at Methodist Hospital.

Hoshin Objective: The Hoshin strategic initiatives adopted by Methodist Hospital prior to the joint venture have evolved from their first five-year plan (Figure 6).

This plan is reviewed yearly. Their changing environment and customer demands keep the plan dynamic. The first Hoshin initiative was to implement Total Quality Management. This was determined to be the key strategy—or driver—to make the implementation of the five-year plan successful. In the second year, through an environmental scan, the growth of managed care in their service areas played a key role in formulating a second Hoshin initiative—Hospital and Physician Relations. Hoshin

Background on Methodist Healthcare System, continued

Planning was even used to develop the original key driver that led to the joint venture with Columbia/HCA Healthcare Corporation.

Figure 6. Five Year Hoshin/Strategic Plan Deployment

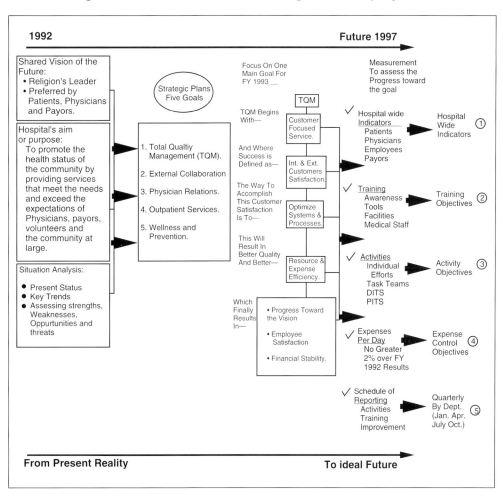

The planning model also integrates a "catchball" communications process

There's also a Catch-Ball process throughout the organization as to how everyone can be involved in that process (see Figure 7). The interactivity of what's been done shows up in a very structured communication process, with "this is what we want to accomplish," "what can you do?" This goes back and forth with the management to be sure that everyone is adopting plans that do accomplish the alignment of their Hoshin.

Annual review and plan update

Because all strategic plans are dynamic—especially in health care—an audit and review process is necessary on a yearly basis (customer demands and environmental trends do change). In the first year of the Hoshin Planning process it was necessary to concentrate on process improvement (TQM). During audit and review, in year two, process improvement was still important but, additionally, the environment and customer demands were changing (managed care entered the market). The resulting review and planning exercises culminated in the adoption of a second Hoshin—Hospital and

Annual review and plan update, continued

Physician Relations—which was deemed to be just as important as process improvement. The TQM driver *catchball* played an important role in this review. During *catchball*, Methodist management was told by their employees to keep the focus on TQM as the first Hoshin but also adopt a second Hoshin. This taught us that it is possible to work on more than one Hoshin at a time.

Methodist /Columbia planning model

Figure 7. Planning Model for the Interim Plan. This Model Was Used During the Start-up Phase of the Methodist/Columbia Joint Venture

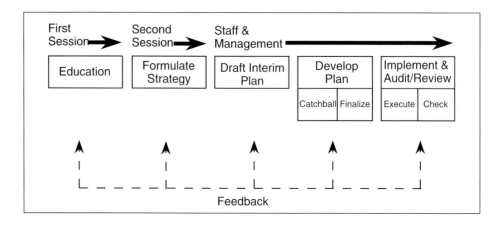

Identifying targets and key drivers

Another important step in this planning process is identifying targets and drivers. The Tree Diagram (Figure 8) is an example of how the plan "rolls out" from the mission, vision and values of the organization through the main initiatives, first-year targets (goals), the Hoshin, the first-year targets (goals of that Hoshin) and the tactics for implementation.

Figure 8. Strategic Plan With Targets And Key Drivers

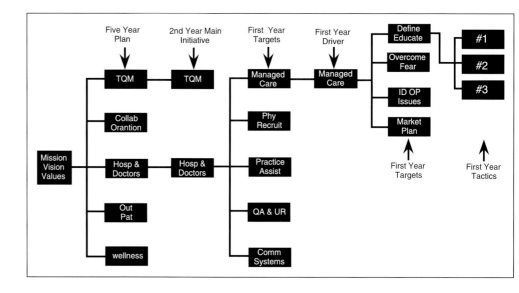

A matrix diagram makes the plan easier to see and understand

An implementation matrix (Figure 9) can be used to visually communicate how every part of the process is coordinated to implement the tactics associated with the Hoshin. Strategies (or tactics depending on the audience using the matrix) are delineated, as are the resources, responsible parties, the metrics, and the development time line. This visualization promotes aggressive implementation and serves as an on-line audit tool.

Although this example is a matrix used at the plan development level, and subject to change prior to catchball, matrices that are further cascaded down the ranks become a valuable tool for alignment of all interests and employee empowerment.

Figure 9. Implementation Matrix

Strategy	Resources	Responsible Parties	Measurements	FY93 Q1	Q2	Q3	Q4	FY94 Q1	Q2	Q3	Q4	FY95 Q1	Q2	Q3	Q4
Creates an Attitude of Collaboration and to do What is "Best" for the Times	Literature Conferences Discussion	PO Board Board	PO Acceptance = 300			x	x	x	x	x	x	x	x	x	
Establish & Implement Physician Membership Process/Criteria	Board Other "IPA" Models Employer Demands	PO Membership Comm	Document with Physician Consensus	x	x										
Sell the Concept to the Consumers (not the only ones)	Methodist Hospital Marketing Dept./ Business Dev. Physician Organization	PO Marketing Comm	# of Customers (trend)			x	x	x	x	x	x	x	x	x	
Help Create "Coordination" Structure	Administrative Input Other IPA Models	Board of Directors PO Comm	Structure Completed	x	x	x	x	x	x	x	x	x	x	x	x
Educate Physicians to Utilize Cost Effective Process	Literature Hands - On Seminars Feedback on Outcomes	PO QA/ UR Comm	Data provided Knowledge about cost		x	x	x	x	x	x	x	x	x	x	
Establish Appropriate Reimbursement & Incentives for Physicians	Community Norms Outcome Measures	Consultant Actuary Questionnaire PO UR Comm	Maintenance of Membership		x	x	x	x	x	x	x	x	x	x	
Establish Outcomes Measurements for Processes Improvement	Other IPA & HMO Models Internal QA/ UR Data Methodist Hospital	Quarterly PO meeting with members PO QA Comm	Bottom Line Employer/ Payor Satisfaction		x	x	x	x	x	x	x	x	x	x	

Hoshin complements TQM, it doesn't replace it

One of the major points learned by Methodist Hospital in their use of the Hoshin Planning process is that it complements virtually all Total Quality Management processes and learning. The tendency in any changing environment is to revert back to comfort positions still entrenched in daily activity. This certainly could have been the case with the changes coming from their joint venture activities. Through the continued use of customer-focused planning techniques, quality was still king. No matter the changes, this planning focus kept alive the concept of customers defining quality and the corresponding activities necessary to exceed expectations, regardless of change. Through all of the joint venture discussions quality was the focal point for both parties, and the tools of Hoshin Planning were used to bring this focus into the forefront. It became a significant driver in the integration process.

Hoshin assists in integration of units in a new system

Methodist Hospital, during the joint venture process, had a very strong desire to maintain its focus on Total Quality Management, as did Columbia/HCA. However, cultures do tend to change during a merger process (one becomes subservient to another), so it was important to both organizations to focus on quality as a main cultural driver during the joint venture process.

The principles of Hoshin Planning and the adoption of breakthrough strategies helped in the integration of the various identities within the new joint venture partnership. The model for developing the overall integration plan is seen on page 40 (Figure 7), and was the beginning point of the alignment of all interests, specifically from a Total Quality Management perspective. Education and strategy formulation by all participants within the planning process (this, in and of itself, creates a common ground for discussions on quality), coupled with catchball (feedback) from all constituencies, ultimately produced the appropriate breakthroughs that drove integration implementation.

The system-wide strategic plan then became the template for each facility's strategic planning efforts. Further examination of customer demands and environmental trends at the facility level introduced facility-specific strategies that continued the process of alignment of common interests (the need for the development of a seamless continuum of care) throughout the system.

Interrelationship Digraph helps to sort out the "drivers" and the "outcomes" among a group of priorities

We've included two of the *Seven Management & Planning Tools* that were used in the Hoshin Planning processes leading to the completion of the joint venture. A number of the Seven Management and Planning Tools are used in the planning process. Two of several tools used during the joint venture planning are reviewed below (Figure 10 and 11). The first tool used after the Affinity Diagram exercise was the Interrelationship Digraph (Figure 10). This tool helped planning team members understand strategy relationships, and to ultimately prioritize those strategies critical to the success of the joint venture.

When working with this tool, the strategies are placed toward the outside of a flip chart. (The strategies are the result of a Brainstorm/Affinity Diagram exercise. They are the Affinity Diagram "headers" that were generated by addressing this question: In order to be thriving in 3–5 years, what will we have to do to amplify our strengths, minimize/eliminate our weakness (seize opportunities, and eliminate/minimize threats?) To complete the ID, look at each of the strategies and determine their relationship with each other, then draw an arrow to the strategy, based on whether it causes or influences the relationship. If there is not a direct cause and effect relationship between the two, don't draw a line—leave it open. To interpret the ID, the strategies with the most arrows out represent the "drivers" or causes of those strategies that have the most arrows in, which are the "outcomes."

As you view Figure 10, it is obvious what strategy has more of an effect on all others and thus is a "driver:" establishing a seamless continuum of care. It has 10 arrows going out and this became the most important strategy for the launch of the Methodist/

Interrelationship Digraph analysis identifies drivers and outcomes, continued

Columbia joint venture. This example demonstrates one of the strengths of using the *Seven Management and Planning Tools* in the Hoshin process.

Figure 10. Interrelationship Digraph: Prioritization and Identification of Which Strategies Are in Service to Another

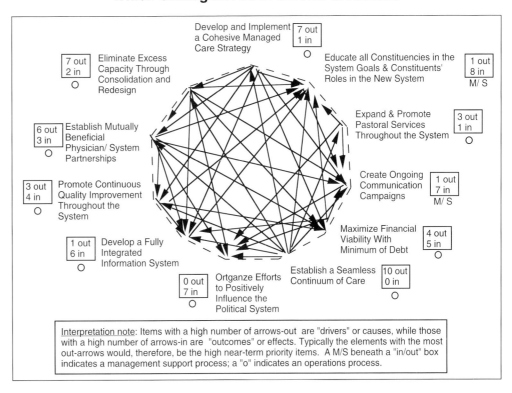

Interpretation note: Items with a high number of arrows-out are "drivers" or causes, while those with a high number of arrows-in are "outcomes" or effects. Typically the elements with the most out-arrows would, therefore, be the high near-term priority items. A M/S beneath a "in/out" box indicates a management support process; a "o" indicates an operations process.

A Radar Chart lets us array, compare, and easily see individual strengths and weaknesses plus relative balance among a group of variables

Another important tool to use in conjunction with the Interrelationship Digraph exercise is the Radar Chart, also called the spider web chart, or the old fashioned gap analysis (Figure 11). This exercise allows the planning process to check the prioritization of the strategies by determining where the organization is today with their strategy implementation versus where they want to be in five years. The gap between the outside edge of the chart and the point along the line from the strategy to the center helps to check and clarify the priorities noted in the Interrelationship Digraph exercise. However, keep in mind both of these exercises are very subjective and are not cast in stone. Ultimate prioritization comes from a keen understanding of the environment and customer demands as they relate to a future desired state. Both of these exercises help planning teams with their decisions—they do not make the prioritization decisions for them.

Discussion—Hoshin Planning creates organizational alignment

Dona Hotopp, GOALQPC—Geoffrey Crabtree, you were talking about Alignment at Methodist Healthcare System. Is that something you're doing? How are you going to use Hoshin Planning for the Alignment that you need going forward?

Radar Chart

Figure 11. Radar Chart of Strategies (Gap Analysis)

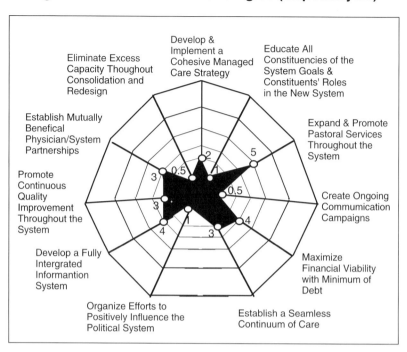

Discussion—Hoshin Planning creates organizational alignment, continued

Geoffrey Crabtree, Methodist Healthcare System—The Hoshin Planning process in and of itself creates alignment. Customer focused planning, to be effective, must include everyone in the process—internal and external customers alike. We create alignment by knowing the demands of our customers and then sharing them, along with the environmental analysis, with our internal customers. This catchball process is perhaps the most important part of this planning rigor because without this understanding of the plan from top-down and bottom-up, implementation does not occur efficiently or effectively—thus alignment does not occur as well.

This is particularly important in a new organization like the Methodist/Columbia joint venture. The success of the negotiations, and finally the partnership itself, can be traced to everyone understanding the mission, vision, and values of the new organization as it begins. When you add the strategies, developed from a clear understanding of all customer demands, to this understanding of mission, vision, and values, you come up with a very proactive partnership with everyone moving in the same direction. For me, that is the power and necessity of alignment.

Finally, the ultimate measure of this success will be seen in how fast the joint venture strategies are implemented without territorial issues and personal agendas acting as barriers. Alignment of interests, as developed through the Hoshin process, has the greatest chance for removing these barriers and then propelling the joint venture to success faster than with any other planning system.

The Hoshin process facilitated the joint venture

GOAL/QPC—Could you talk a little bit about what you're doing for Alignment in your new joint venture?

Crabtree—We began promoting the alignment of interests even before the joint

The Hoshin process facilitated the joint venture, continued

venture was finalized. We began this alignment through a joint strategic planning process approximately four months prior to the closing. We did this because of the unique aspects of the joint venture and the desire to quickly involve both organizations in the understanding of this different partnership.

Essentially, each partner placed their assets into a new partnership to be co-equally owned and managed. Neither organization aspired to sell or buy any of its current assets. By combining these assets, a true 50/50 partnership was created, with consensus necessary on all issues. With the necessity to arrive at consensus on every issue, one can readily see the need to proactively create alignment of all interests and understandings as soon as possible. The Hoshin planning process helped both partners with that journey.

The actual catchball process of the partnership strategic plan, begun during the final partnership negotiations, proved to be a unique experience for all employees. We couldn't think of a better way to foster alignment of a truly powerful partnership than through presenting the strategic plan for this partnership throughout the organization as quickly as possible. A bound booklet describing the work product from the joint planning retreat was published and given to all of the leaders of each facility within the partnership. This booklet introduced the mission, vision, and values of the new organization; a SWOT analysis; environmental trends; customer demands; the strategic initiatives; and an implementation matrix. At the end of this booklet, a very specific catchball questionnaire was developed for review by all employees. The individual facility managers were asked to review this work product with their employees and to solicit feedback based on the questionnaire. This feedback was then sent to the partnership's senior management for review. Appropriate changes were made to the plan, objectives and tactics were added and then sent back to each facility for round two of catchball—with a different presentation book and questionnaire.

Strategic plan is a platform for alignment

Simply put, alignment of interests begins with the simple task of asking everyone their thoughts about the organization's mission, vision, values, and strategies—before implementation. In this environment, the Hoshin does not become a weapon, but instead, clearly-defined strategic initiatives with everyone knowing their role in its implementation. To accomplish this within a significant amount of change, and all its accompanying baggage, is very important as you start-up a unique partnership. Our success in alignment of interests isn't total, but without this type of planning it is doubtful the partnership would be moving as fast in the market as it is.

Hoshin Planning is used for cost reduction

GOAL/QPC—So we see that Hoshin Planning helps with Alignment. One of the main issues facing us every day is cost reduction. Owen McNally, please tell us a little bit about your accomplishments at Our Lady of Lourdes Medical Center in New Jersey, how you're using Hoshin for cost reduction.

Owen McNally, Our Lady of Lourdes Medical Center—We started Hoshin Planning at Our Lady of Lourdes Medical Center three years ago. The first year was

Hoshin Planning is used for cost reduction,
continued

spent simply in planning. We actually started to use Hoshin Planning to align the organization around our major objective, which had implications of cost reduction, but which was broadly directed at reducing clinical utilization and improving quality monitoring.

What we were measuring, in order to see whether we were effective, was the process time or delivery component of "reduction of length of stay." The Japanese, in their Hoshin plans, kind of look at four major areas: quality, cost, delivery, and employee morale. We were really only measuring one of those—delivery process. But it did have implications for our cost reduction goal.

There are specific elements of our plan where we were able to show (for instance, our lab project) a reduced utilization of "chem 12's" and "chem 20's," from 48,000 to 28,000. There was over a $100,000 savings just in that one piece. But one of the difficulties that we have, which I know other institutions have, is that we do not have the information systems in place to really measure the cost of what we're doing. That's something we're going to change quickly. Our annual budget is now $130 million. Before 1993 we generally showed a bottom line surplus of about a million dollars. Maybe, if we had a good year, $2 million. In 1993 utilization and length of stay reduction efforts generated $10 million, which were reinvested in the organization. However, with the move toward Managed Care these results will be difficult to replicate.

When we first started this project we were using *Reduction of Clinical Utilization* as our Hoshin. We looked at a report that the Governance Committee produced, which predicted that in a 400 bed hospital a reduction in length of stay of one day would probably have about a $4 million impact on the bottom line.

In 1993 we saw a $10 million impact on the bottom line. While some of it can be attributed to external influences of Managed Care, that alone could not have produced the magnitude of financial improvement that was realized. When we started, average length of stay was 9.3 days. By June of 1994 it was down to 7.3 days.

When we first started in Cardiology Services average length of stay was 9.4 days. In the third quarter of 1994 it was 6.56 days. It has a significant positive effect on the bottom line when we are able to move more patients through. As you can see in Figure 4, in each quarter we've been improving. Basically, the line has been going up in terms of the number of patients that we've been able to provide services for, and the *Length of Stay* continues to drop. In the first quarter of 1993 we showed a total of non-inpatient heart catherizations of 43. By the third quarter of 1994 it was 228. Our cardiac specialties are our major tertiary service. In 1993 we performed close to a thousand open-heart operations and over 3000 catheterizations and angioplasties. As we become more efficient we're able to help more patients, while at the same time improving the bottom line.

Question on patient care vis a vis shorter stays

Q: How do you make sure that patients are going home in better shape in fewer days?

Question on patient care vis a vis shorter stays, continued

McNally—That's a critical issue in terms of what's happening in health care today, because the pressure is on to move patients through the system as quickly as possible. Our Utilization Management people are committed to Quality. We try to be a high quality, low cost provider, so there are times that we don't send people home on the days that the Managed Care companies want them to go home, because we don't feel that they're ready to go home at that particular point. With that philosophy in mind we still find that we're able to send people home many times feeling better than they did before, because of the improved techniques that are being used.

While this is just an overview, one of the things that has helped us is that, in our Cardiac Services, we've been able, on a quarterly basis, to provide all of our cardiologists data specific to their own practice and specific to the DRG's that they're working on. This helps them to see the type of quality that they're providing compared to the best practices within the hospital and the best practices within the state.

As we look into the development of our 1995 Strategic Planning process, of which this element will continue to be a piece of it, we're going to be using our Hoshin process to help us develop the operational objectives that will move us toward the creation of the integrated delivery network. Next year we're going to be able to show not only the process time measures that we are using, but also the Quality, Cost, and Customer Satisfaction measures.

Question on using data for improvements

Q: Are you gathering data regarding fewer instances of staph infections, post-op recovery periods, changes to less invasive procedures?

McNally—Yes, these improvement efforts are going on. In Figure 3 you saw that one of the Hoshin initiatives, along with the work being done in cardiac surgery had to do with the reduction of hospital infections. There is more data to support these initiatives at our medical center which is not shown in this paper.

Crabtree—I think if you look again at the Fishbone Diagram (see Figure 3) you'll have a sense that it's just not clinical. There are so many different areas that are involved in *Reduction of Length of Stay*, and of course each one of these areas has data unto itself that would need to be examined. The work the Cleveland Clinic has done is a perfect example of reducing length of stay. They have a two-hour presentation on how they reduced Length of Stay in their Cardiac Department, looking at each one of these individual areas as potential for improvements.

Question about the impact on readmissions

Q: When you reduce length of stay do you increase readmission rates?

McNally—We do measure that and we have been seeing a reduction in our readmission rates, too.

Question about employee empowerment

Q: What about Empowerment through the Hoshin process? Everyone talks about empowerment, relationships, information. How has the Hoshin helped you develop the relationships, give people the information they need, and let people have a sense of what they're doing, self-reference?

Question about employee empowerment, continued

Crabtree—Empowerment is a real interesting philosophy or concept for me, because empowerment would imply that someone gives employees the right to think for themselves, to take risks, and to do all the things that you'd hope your employees would do. But you also add two other dimensions to that, and that is the dimension of whether the employee wants to be empowered, and whether the employee actually feels empowered. Those are three dimensions that work hand-in-glove and seemingly are, at least for us, solved by the Hoshin Planning process.

When you ask people to get involved that begins the empowerment process. But just as importantly you've got to react to their comments back to you. When you react to their comments back to you, when you change the Strategic Plan, for instance, or change a part of a tactic, time-line, resource, responsibility, or even a measurement, when that's changed because your employee has indicated something to you that's very important, then that employee feels encouraged, at least, to ask or suggest again. That is empowerment. People become empowered more from the reaction to their suggestions back to you than from anything else.

The next part of it is, how do you talk to employees? How do you get employees who don't want to be empowered, empowered? There are some employees who simply want to come to work, do their job, and go home, and offer absolutely no suggestion on how to improve their job or anything. Oftentimes those people have important items to suggest. They are people that you really want to empower to speak their mind, empower to take risks, empower to offer suggestions for improvement.

I think the Hoshin Planning process or Customer Focused planning process, as we call it, encourages that kind of empowerment or that kind of activity to empower. It does so, not from the fact that management is asking for your input, but because there is a system that has been put in place to listen to the customer.

When the employee hears the customer talk about what they expect, want, and need, then that customer is really saying to that employee, "Help me get what I need. Do those things that will make me happy." That in itself is empowerment. So it's the system that is put in place by Customer Focused planning that helps empower people.

That system is embodied by the Catch-Ball process, simply asking people their opinions about the Strategic Plan, about objectives and tactics, about resources and responsibilities and measurement, about the time-line for accomplishment. It also sets up the system that acknowledges the customer's point of view, because the employees, the staff, those folks closest to the customers should have that data available. Customer Focused planning is one of the most important empowerment tools I know of.

A formal review process moves the plan along toward achievement and improvement

GOAL/QPC—Let's talk about the Review. One of the issues that's probably most foreign to us as we do a planning process is the Review process, in terms of maybe we've never gotten far enough to do a Review process. Owen, what are you doing to try to introduce that Review process, and how are you going to improve it?

McNally—The Review process is a critical element in trying to understand

A formal review process moves the plan along toward achievement and improvement, continued

better what Hoshin Planning is all about. There are two major elements, as I see it, to the Hoshin Planning process. One is as Geoffrey mentioned—it builds Quality into the traditional Strategic Planning process. Secondly, it is the organization of the review process that has enabled us to do is get the Strategic Plan off the shelf.

We meet on a monthly basis with all of those people individually who had been involved with the Hoshin Planning initiatives. We've tried to keep them focused on their efforts. We're learning as we go along, so the only measurement we've been working on has been the reduction of process time, either Reduction of Length of Stay or Reducing Clinical Utilization.

In 1995 when we meet with the department heads and clinical chiefs on a monthly basis we are asking each initiative leader to determine up-front what will be their Quality, Cost, Delivery, and Customer Satisfaction targets. And we'll be monitoring those with them on a monthly basis. This review will enable us to make adjustments as we move along. If we're not meeting targets, ask the question "why not?" and "are there any changes in the plan that have to be made," rather than wait until the end of the year to do that.

One of the other things we did that we found to be very helpful is we had a quarterly Hoshin presentation meeting where 50-60 of the department heads from the medical center would come together, and each person who was responsible for a Hoshin initiative would get up and in five minutes give an overview of the work they were doing, either with a story board or with graphs and charts. That enabled everyone within the medical center to realize that "what I'm doing" is not being done in isolation; there are others who are doing something different but we're all contributing to the vision of the organization. We found both of those review processes very helpful to us and we'll continue to refine and improve them. It's been a good start.

Question about physician involvement in the process

Q: Explain how you involve the physicians.

McNally—When we first started Quality management in our institution with project teams we'd hoped to have a lot of physician involvement. We found that very difficult. The percentage of physician involvement at that time was minimal. When we started our Hoshin Planning process, being aware of the negative difficulties that we had with the process teams, we decided up front that we had to get our physician leaders involved.

When we went through this whole planning process, both our planning workshops and weekly team sessions to develop the tree diagrams, our medical chiefs were involved almost 100% of the time. We had fifteen of our medical chiefs who participated in this process and they stayed with it right to the end. They understood what we were doing and why we were doing it. It also helped to bring about a certain collaboration between our administration and our physicians. In most hospitals, and ours is no different, they often seemed to be going along on two different tracks. Administration gets involved in the planning and most of the physicians don't really know what's

Question about physician involvement in the process, continued

happening.

We found this joint physician and administration involvement to be very helpful. We achieved 100% buy-in with everything we are doing. For instance, a lot of the improvement in our cardiac services, which is our largest tertiary service, is really being led by physicians.

Question about joint venture relationships

Q: Mr. Crabtree, you mentioned the joint venture partnership. Have relationships changed? Have they been won over to a real belief in the way you are doing things, and if not, how is that effecting the outcome?

Crabtree—It's not a question of winning anyone over. A true partnership is never us versus them. It is more the reality of setting up the mechanisms and environment that fosters consensus, alignment of interests, and empowerment of all employees. The new partnership acknowledged that the mission, vision, values, and policy on quality of the Methodist organization would be adopted for the new partnership. The Social Principles of the United Methodist Church, with its position on mission and quality, was also adopted as an important foundation for the partnership.

These documents formed the base to begin the Hoshin Planning process. We all said this was where we were going to begin to promote alignment, so very necessary in a new partnership. However, this was just the beginning point. Quality was and is still the overriding issue. The power of the concept of Hoshin Planning did more to bring a common understanding of quality to the table than anything else—and for a simple reason. Customers define quality. If you adopt a planning process that is customer focused, internally and externally, you automatically build quality into your short and long-term strategies. To bring together two organizations that were strong competitors and get them to think alike is a vary daunting task. When you do it from the customer's point of view—the remarkable, yet simple concept behind Hoshin Planning—this task is made easier. More importantly, the outcome is powerful in the market, especially if accomplished in a very short period of time.

It is a wonderful experience to ask what the customer wants. We believe strongly in the power of this type of planning. To begin a partnership, and the necessary strategies that build quality into every process, with anything less than a customer-focused process in today's constantly changing environment portends a long and difficult journey. After all, the market is the driver, and customers—internal and external alike—make up that market.

Author information

Geoffrey Crabtree is Vice President, Strategic Planning and Market Services for Methodist Healthcare System in San Antonio, Texas. He joined that organization in 1988 as Director of Marketing. Prior to that he held many marketing and strategic planning positions in various companies. He is the founder and past president of the San Antonio Chapter of the Academy of Health Service Marketing, and has written and published several articles on health care marketing and public relations, and has been awarded over 50 local

Author information,
continued

and national awards for health care planning and strategic marketing.

Dona Hotopp is GOAL/QPC's Director of Health Care Services, director of the GOAL/QPC Health Care Application Research Committee, and a member of the planning team for the Community-wide Health Improvement Learning Collaborative. Dona is coeditor and coauthor of Putting The "T" In Health Care TQM, and GOAL/QPC's Integrated Planning Research Report.

Owen McNally joined Our Lady of Lourdes Medical Center in 1990 as Director of Total Quality Management. He was previously Executive Vice President of Marc Associates, a consulting engineering firm in Mt. Holly, NJ. Prior to that he was Township Manager of Lacey Township in Forked River, NJ, and Administrator of the St. Francis Community Center in New Jersey. Mr. McNally serves on the Advisory Board of PACE, the Philadelphia Area Council for Excellence.

Managing by Teams at Rush Home Care Network:
It's not the Function, it's the Process

Author

Kathryn E. Christiansen, D.N.Sc. Executive Director & Administrator, Rush Home Care Network, Rush-Presbyterian-St. Luke's Medical Center, Chicago, Illinois

Introduction

Most people within the health care industry, and particularly with home health care, tend to believe that we are facing the toughest problems and everybody else has it easy. However, the reality is that anyone in business, regardless of size, location or industry, is experiencing similar problems. This article will describe how my particular organization, Rush Home Care Network, has dealt with the issues of moving an organization forward in a turbulent health care environment.

Background & Environment

A description of Rush Home Care

Rush Home Care Network (RHCN) is a hospital-based agency within the organizational structure of Rush-Presbyterian-St. Luke's Medical Center. The hospital, a large tertiary referral center with associated colleges of medicine, nursing and health sciences (Rush University), is located in the Chicago loop area. Rush Home Care Network, originally initiated as a training site for Rush College of Nursing students, has been in business for over 22 years. RHCN provides intermittent nursing, physical therapy, occupational therapy, speech pathology, medical social work, and home health aide services to patients who are home-bound. During the last fiscal year, RHCN provided nearly 150,000 visits to nearly 4,500 patients. At any given time RHCN has around 1,000 patients in service. Home care services must be medically necessary, related to a primary diagnosis and authorized by the patient's physician. Since the enactment of Medicare in 1966, home care has become more widely known and used. Literally all insurances provide home care coverage. The largest home care payor, by far, is Medicare (70 - 90% nationwide) followed by indemnity plans, Medicaid, HMOs, PPOs, workers' compensation, and private pay.

Reacting to changes in the health care industry

With the advent of Medicare reform in 1984 and prospective payment, hospitals began using diagnostic related groups (DRGs) to reduce lengths of stay and allow patients to safely continue their treatment regime at home. Home care was the obvious solution to provide a seamless transition from hospital to home. Post-acute treatment in the home was preferred by most patients and assured physicians that observation and

Reacting to changes in the health care industry,
continued

treatment would continue until the patient had recovered or stabilized. Home care use and expenditures began to grow as a result of several factors: the needs of an aging population; a response to "quicker and sicker" hospital discharges; the demand from managed care organizations to treat patients as outpatients whenever possible; and changes in technology allowing more complicated procedures and treatments to be done in homes. The Health Care Financing Administration (HCFA), now viewing home care as an "industry out of control," began testing a prospective payment methodology as a way to reduce or stabilize expenditures. In addition, managed care organizations aggressively negotiated discounted, per visit, and capitated rates far lower than Medicare's caps. Home care agencies started to consider their current cost of doing business as a weakness and a threat. Rush Home Care Network, like most other home care agencies, looked for ways to simultaneously reduce cost, improve quality and improve access to care.

Care Teams

The discovery of care teams as a possible solution

The idea of clinical teams was proposed after two nurse directors attended a two-day conference on "Ideal Care Teams" offered by Larry Campbell, author of *The Ideal Care Team Model for Home Care*, in 1993. Several home care agencies discussed their experiences with teams that were largely composed of nurses, led by nurses, and focused on nursing issues. The attending directors were favorably impressed, believed the concept could be expanded to include all disciplines, and presented the idea at a Directors' Meeting. There was initial interest and support of the idea as a strategy to streamline work and reduce cost. However, that enthusiasm was soon followed by hesitation and doubt that care teams could work for the multiple disciplines employed at RHCN.

Interdisciplinary care teams

Our agency administrator had marveled at the efficiency of small, simple home care organizations of about 50 to 100 patients. Employees in these agencies were intimately involved with all aspects of the operation and were familiar with the needs and treatments of every patient. Five years earlier, the idea of breaking the large agency into a series of small ones had been presented to RHCN staff. At the time, we were going through an organizational change, and the idea was summarily dismissed. Instead, we implemented the existing organizational structure, with multiple clinical directors each responsible for a single discipline and managing a cost center.

The implementation of interdisciplinary care teams was debated for more than six months. The questions were numerous—people raised concerns over potential problems, and were worried about changing a structure that they thought was working. However, our costs of business were too high for us to survive in a managed care environment.

Some reasons for rethinking processes

State and JCAHO (Joint Commission on Accreditation of Healthcare Organizations) surveys consistently showed a lack of documented interdisciplinary communication and care coordination. Numerous attempts to encourage timely submission of paperwork, enhance the content of documentation, and improve billing processes only resulted in marginal improvements. We realized that we had developed our systems in response to regulation, adding non-value added steps to our processes. We had too many hand-offs, checks and balances, and redundancy.

Communication with our staff was poor. They were located throughout Chicago delivering care to patients in their homes, and had difficulty contacting us or other staff members in the field.

The team model we were considering as a solution consisted of small groups of people delivering care, which was primarily nursing driven. In our model, quality would be defined by the customer.

A decision to charter two pilot teams

After managers talked through all their concerns, only two major, unresolved issues remained: 1) Does the team leader need to be a nurse?; and 2) Can billing be done at the team level? The decision was made to pilot the model for about four months with two interdisciplinary care teams; one in the Chicago office and one in the North Shore office. At the end of that time, if results were favorable, we would implement the teams throughout the whole organization. They would be designed to serve the average home care needs of 120 to 150 patients in a specified geographic area. The makeup of the pilot teams is described in Figure 1 below.

Figure 1. A Description of the Pilot Team

- One team leader—expected to spend half time with patients and half time as a facilitator of team processes. A nurse served as the team leader for the pilot.
- Three medical-surgical nurses.
- Two psychiatric nurses with the ability to see uncomplicated medical-surgical patients.
- Two physical therapists plus a physical therapy assistant (P.T.A.).
- Two home health aides.
- One occupational therapist.
- One social worker.
- Speech pathologist as needed.
- Part-time clinical staff, as needed for high volume.
- Two team assistants (one clerical support, one biller).

Defining director responsibilities

We identified geographic boundaries that the teams would work within, so that they would be closer to their patients. We solicited clinical volunteers and clerical support staff, and the study questions were narrowed down to a reasonable number. Director involvement was designated as follows:

- One director served as coach in the Chicago office.
- One director served as coach in the North Shore office.
- Two directors agreed to plan and implement the educational and training modules.
- One director worked with the team assistants.
- One director sorted out team-specific data and generated reports.
- Three remaining directors helped out as needed.

Components of team focus

The model for team focus and educational programming as designated in the Joiner Triangle (1996) consisted of: Education to quality and customer value; A scientific approach to continuous improvement, and; An environment of teamwork and cooperation. Quality would be defined by customers, both internal and external. Internal customers were other members of the team and the agency as a whole. External customers were defined as patients and families, physicians and discharge planners, and regulatory bodies such as the State of Illinois licensing criteria, Medicare standards of participation and JCAHO (Joint Commission on Accreditation of Healthcare Organizations) accreditation standards. All One Team was defined as team-building and the spirit of cooperation and teamwork that would occur throughout the pilot process. Scientific Approach was defined as the process improvement activities, using data to make decisions about patient care, and the evaluation of the pilot.

Asking questions of ourselves

We asked some questions of ourselves to try to understand how we were doing:

1. What is the right size and composition for interdisciplinary care teams?
2. Does the team leader need to be a nurse?
3. How much initial and ongoing team education is needed?
4. How frequently should the teams meet?
5. What does "self-directed" really mean?
6. What functions need to remain centralized?
7. How will managers function?
8. How should we measure success or failure?
9. How can incentives be team focused?
10. Do we have improved systems?
11. Do our customers like the change?
12. Do we have better business results?

Measurements of improvement and data collection

We set different measures of process improvement and methods of data collection:

1. Fewer average visits per patient (statistics)
2. Shorter average lengths of stay (statistics)
3. Easier and better communication among disciplines (documentation, observed interaction, and reports by discipline)
4. More interdisciplinary referrals and better mix of services per patient (statistics and observation)
5. Better compliance with standards of timeliness (quality monitoring)
6. Reduced cellular phone and credit card usage (check phone bills)
7. Lower cost per visit (cost/visit reports)
8. Better charge capturing (suspense report, and billing personnel report)
9. Same or better patient satisfaction (satisfaction surveys)
10. Increased staff satisfaction (opinion surveys, pilot team interviews).

Two pilot teams at the forming, storming, and norming stages

The Chicago Pioneer Care Team was formed in May, 1995, and the North Shore Pioneer Care Team formed mid-June, 1995. The forming stage varied with the two teams.

The Chicago Pioneer Care Team formed quickly and moved right into the Storming phase, which lasted throughout the summer. The team leader described one experience in the following way:

"Nothing made the therapists happy. We started out by meeting every day. Physical and occupational therapists complained that they didn't want to come into the office that often. We decided to meet in the community so team members didn't have to come into the office. That didn't make them happy. We altered the time of day. That didn't make them happy. We finally cut back to three times a week. That still didn't make them happy. Finally, after dealing with this unhappiness all summer, I came to a team meeting and said, 'I am not enjoying this experience! It's too hard! I'm not having a good time!' Everyone at the meeting said, 'We're not enjoying this either. It is too hard.' Then someone on the team said, "What are we going to do about it?" At that point we started to move into the norming phase."

The North Shore Pioneer Care Team spent most of the summer forming. They did a lot of group processing but not very much care coordination. As a result, team members liked each other and rarely disagreed. Patient care was not often discussed during team meetings. The team decided early on to meet three times a week, and later reduced it to twice a week. While patient care occurred, it differed very little from the more functional approach to care in the office. This team didn't storm throughout the first four months of the pilot. Later, the team leader left the organization during an organizational consolidation. That's when the team began to storm and shortly thereafter moved into a norming phase.

Performance after four months

Everything was moving along more or less as we had anticipated. After four months, we collected data on length of stay, visits per patient, and standards of timeliness (see Figure 2). Length of stay is used to describe the average number of days from a patient's admission until discharge. This is a good measure of our effectiveness in achieving outpatient-specific outcomes. The process followed by the Pioneer Care Teams was to discuss the plan of care, agree on the role each discipline would play, evaluate ongoing progress toward goals, and discuss the outcomes prior to discharge. The remainder of the organization continued to approach care from an individual discipline perspective. Interdisciplinary case conferences were not consistently done, the plan of care was not unified, and the patient was discharged when individual objectives were achieved. The result was usually a longer average length of stay.

Standards of timeliness are used to describe the amount of time it takes for clinicians to turn in their paperwork. Some examples of paperwork are admission information, physician orders, visit notes, paraprofessional supervision, and discharge paperwork. The standard for the agency is 3.5 days, and enforcement of that standard has been an ongoing challenge.

Figure 2. Data Collected after Four Months

	Pioneer Care Teams	Agency
Length of Stay[1]	95 days	122 days
Visits per Patient	31.87	43.62
Standards of Timeliness[2]	4.1 days	4.6 days

1. Length of Stay refers to the average number of days from the patient's admission to the home care agency until discharge from the agency.

2. Standards of Timeliness refers to the amount of time it takes for clinicians to turn in their paperwork.

With minimal coaching, the Chicago Pioneer Care Team appeared to be making 80% of the day-to-day patient care decisions. Directors were encouraged by the results. It was time to make a decision to continue or stop. No decision was forthcoming. Directors resumed their earlier hesitation and doubt. Old questions, like whether the team leader should be a nurse, resurfaced and were debated. The question of cost was raised: Was this really a less costly model?; How many team assistants could we afford to have?; Wouldn't we actually need more people? We decided to reduce team assistants to one per team, to determine if the work could be done by one person. This new test would require at least two months.

Better performance and more results after 10 months

We evaluated the data after 10 months:

- Average standards of timeliness continued to be lower for the Pioneer Care Teams

- Charge capturing was better, as reported by payor relations; fewer bills from the Pioneer Care Teams were in suspense

- Documentation reflected better coordination of care among disciplines on Pioneer Care Teams.

- Discipline mix for patients on Pioneer Care Teams was more balanced than for the agency as a whole

- Cellular phone use and credit card calls showed no discernible difference from the rest of the agency

- It was impossible to determine cost per visit by team because of the way we captured for the agency as a whole—after considerable effort, attempts to provide this measure were abandoned

- Some patients called to say they liked the way we were delivering care. They commented on how clinical staff informed them about the visits that would be made by other members of the team. One patient said, "I used to think that your staff never spoke to one another, only to me. I was asked by the nurse when the physical therapist and social worker were coming to visit me and I was expected to keep track of notes that one person would leave for another. It seemed like I was the only one who knew what was going on." The wife of another patient complimented our coordination of care and sent a $500 check for a team party

- Feedback from the Pioneer Care Team members was positive. They liked the model and thought the agency should move forward to full implementation.

Some lingering problems still existed

In spite of compelling evidence that the model was working as expected, the directors continued to debate. Some directors were meeting among themselves to discuss the negative aspects of the model. It was becoming obvious that no progress was going to be made under the current organizational structure. Several authors of organizational redesign have warned against changing only part of an organization. Hammer (1993), in his book *Reengineering the Corporation* states that everything within the company is linked together: business processes, jobs and structure, management and measurement systems, values and beliefs which, in turn influence business processes (p. 80). Champy (1995) in *Reengineering Management*, indicates that changes in an organization must be accompanied or even preceded by changes in management. Our inability to move forward was because we were trying to change business processes without changing jobs and structures, management and measurement systems, and culture. We went back to the drawing board.

Looking at the Baldrige Award Criteria

Leadership made the decision to examine the 1996 Malcolm Baldrige National Quality Award Criteria framework. The framework has three basic elements that form the dynamic relationship among the categories. The "driver" is leadership as the organization defines it. The "system" is made up of process management, human resource development and management, strategic planning and information and evaluation. The "goals" are: (1) customer and marketplace performance, and; (2) business performance. The directors were asked to review the Baldrige criteria and redesign Rush Home Care Network to meet the specifications of the model.

Three groups of directors worked on three designs and presented them. After much discussion, we decided to designate three process areas:

1. Customer Access, or entry into the system,

2. Service Delivery & Service Support, or service to our customers, and

3. Information & Evaluation, or the gathering and analysis of information.

All existing functions would be redesigned and assigned to the process area that captured most of the functional activities. All functions were identified by the director's group, typed onto yellow sticky notes and placed on a large piece of flipchart paper. Functions could be moved around on the paper until the new design made the most sense. After a tentative agreement about the design, which included full implementation of the interdisciplinary care teams across the network, the directors voted on implementation of the model. The vote was seven for implementation and two against, and majority ruled. The administrator, although in favor, did not vote. The two against didn't feel that this design was the right thing to do. Of the seven directors who voted in favor of redesign, most spoke against it in private sessions with one another and word got out that the directors were not in agreement about this model. One director from the work group that designed the model and voted for the change, submitted her resignation within the week. Another director who believed in the interdisciplinary care team model but saw no place for herself in the new design, decided to transfer to a college faculty position. We didn't drop anybody out of the organization, but we changed their jobs and asked them then to interview for those jobs.

Administrator appoints new directors

The administrator assumed responsibility for appointing three directors to design the three process areas. No assumptions were made that these individuals would assume the leadership positions, however, that is what eventually happened. Having agreed to a new model, the next steps were to work on the implementation plan. New and revised positions were identified and job descriptions were written and/or rewritten. Figure 3 on the next page shows the major differences between a functional organization and a process organization.

Administrator appoints new directors, continued

Figure 3. A Functional Organization versus a Process Organization

Functional Organization	Process Organization
1 administrator	1 leader
9 functional directors	3 process directors
8.5 supervisors	4 managers
4.5 clinical specialists	6 clinical resource team members
1 team leader	4.5 team leaders
(0.5 / pilot team)	(0.5 for 9 teams)
12 clerks (all grades)	12 team assistants

Identifying three process areas

The three process areas that were identified were:

Partnership Development and Customer Access *(Knowing the Right Thing to Do).* Includes everything that happens before a patient is admitted into service:

- Marketing

- Hospital and Community Liaisons

- Referral intake, benefit verification and visit authorization

- Reception

- Contract negotiation and management

- New business planning and development

- Strategic planning

- Budget planning.

Service Delivery and Service Support *(Doing the Right Thing).* Includes everything that happens from admission to discharge including services that support service:

- Nine interdisciplinary care teams

- Clinical resource team (advanced practice clinicians, all disciplines)

- District management (three districts)

- Human resources

- Payroll and payables

- Billing and collections.

Information and Evaluation *(Knowing We've Done the Right Things Right).* Includes process improvement activities and the resources to support data capture, report generation, analysis, and education:

- Process improvement team

- Medical records team

- Educational and training support

- Financial systems (reporting, analysis, costing, activity based management)

- Information systems (Maintenance of hardware, software).

Four directors decide to change jobs

As each of these areas were designed and jobs were reconfigured, the remaining four directors evaluated their willingness to assume new jobs. Three decided to leave the organization but willingly helped out with the interviewing, selection, education and training of team members. One director of a new process area also elected to leave the organization after realizing that his philosophy was inconsistent with the new structure.

Assembling and training the seven teams

Team leaders and team assistants applied for the seven new teams. Applicants were interviewed by three directors and the administrator. Part of the interview was used to reinforce the new culture, specifically: customer focus, all-one-team, and process improvement. Successful applicants were notified and training began. Very few of the new jobs were filled with people outside of the organization.

In May, 1996, full day workshops on team building were required for every employee of RHCN. The workshops were facilitated by an individual from hospital human resources. *The Team Memory Jogger*™ was distributed to all attendees and used to reinforce key components of team behavior. During the workshops members of the Pioneer Care Teams described their experiences, both good and bad and responded to questions. A process director or the administrator was present a each of the four workshops to respond to questions, concerns or complaints.

On June 3, 1996 four interdisciplinary care teams were initiated. On July 1, 1996 the remaining three teams formed. On July 1, 1996 (beginning of a new fiscal year) the complete reorganization was fully operational. Ongoing educational programming included: team skills; stages of team growth; working as all-one-team; running successful meetings; conflict management; customer focused behavior. Scientific approach education was scheduled for later in the educational schedule so as not to overwhelm staff with too much, too soon. Process improvement discussions, however, were a part of most discussions with team leaders and team assistants. Team assistants, in an effort to improve their work flow, decided among themselves to break their work up into functional components—"Mary is good at data entry so she can do all the data entry, Jan understands scheduling, so she can do it all, and Sue is good on the phone, so she can take all the phone calls, etc." After our initial surprise that we could so quickly revert back to a functional design, we realized that we needed additional education to reinforce the fundamental philosophy of the model with everyone.

A current status report

We also set about revising all the job descriptions to include core behaviors: customer focus; team skills; and improving processes and core skills: interpersonal skills; problem solving; and self-evaluation in addition to the job components. Staff were informed that a review of core behaviors and skills would count heavily on the merit review. However, in the first year of implementation, staff would rate themselves on the core skills and behaviors, and use the assessment as an educational tool or a plan for

A current status report, continued

improvement during the evaluation process.

A simulated work evaluation was designed for the team assistants to test their skill and accuracy at interpretation and entry of physician orders. While certain weakness were found, all team assistants passed.

Most team leaders are transitioning well. One team leader is having problems but that team isn't helping. The team has asked for outside assistance to help them through some of the *storming*, and human resources has agreed to help. It has been hard to determine where the problems stem from. The district clinical manager has been working with them as a coach, providing feedback to the team leader on approach and style.

The leadership group has also been *storming* on occasion. We don't always see eye-to-eye. We have received feedback from staff that we would benefit from attending the conflict management series. "Put your money where you mouth is!" We are scheduled to follow their suggestion. The administrator has, throughout the year, and continues to hold monthly Town Hall Meetings for any staff that want to come. Because of the size, three meetings are held. Attendance has been good. Questions have also been good. The grapevine is alive and well and the most outlandish rumors circulate. Unfortunately, rumors are rarely brought up at Town Hall Meetings so it is hard to respond. Over all, three months into the new design, transition is pretty much on schedule.

Looking ahead

Rush Home Care Network is on the brink of a culture change. We are not there yet and may not be for some time but the worst thing that can happen is getting "stuck in the middle." I believe that the key contributor to changing our culture is a fervent reinforcement of the components of the model we have chosen to use—we must be customer focused. Everyone must know and understand that everything they do affects someone else (internal or external to the organization) and contributes to our success or lack of success. The all-one-team concept is more than words. It really means that teams are more successful than individuals alone. It also means that one team cannot become insular in their outlook. We must improve our processes. We have to strive to do everything right the first time and not accept second best. We must also remember to blame the process and not the people. That is particularly hard to do. Staff are looking to me and the rest of leadership to set the example, to see if we practice what we preach.

Conclusion

Of all these challenges, the greatest one is to encourage the interdisciplinary care teams and other work teams (e.g. payor relations, referral intake, process improvement) to bond and perform but not to the extent that they become fiefdoms in the process. I hear things like: "We don't do things that way in my area." "No one understands the unique aspects of what we do but us." "I can't allow my people to attend (agency sponsored) programs because we're too busy." I begin to worry when comments like those come from a leader. This is the "make or break" point and I am convinced that a successful future rests on our ability to actualize our team model.

References

Adams, S. The Dilbert Principle. New York: HarperCollins.

1996 Award Criteria (1996) Malcolm Baldrige National Quality Award. Gaithersburg, MD: National Institute of Standards & Technology. (301) 975-2036.

Barker, J. (1992) Future Edge. New York: William Morrow & Co.

Campbell, L. The Ideal Care Team Model for Home Care. American HomeCare Management, in association with the National HomeCare Database.

Champy, J. (1995) Reengineering Management. New York: HarperCollins.

Hammer, M. & Champy, J. (1993) Reengineering the Corporation. New York: HarperCollins.

Joiner, B. (1994) Fourth Generation Management. New York: McGraw-Hill, Inc.

Katzenbach, J. & Smith, D. (1993) The Wisdom of Teams. New York: HarperCollins.

Mears, P. (1994) Healthcare Teams. Delray Beach, Florida: St. Lucie Press.

Scholtes, P. (1996) The Team Handbook (2nd Ed.). Madison, Wisconsin: Joiner Associates, Inc.

Senge, P., et. al. (1994) The Fifth Discipline Fieldbook. New York: Doubleday.

Team Memory Jogger (1995) Madison, Wisconsin: GOAL/QPC & Joiner Associates, Inc.

Author information

Kathryn Christiansen has 32 years of experience in home and community care as a staff nurse, educator of undergraduate and graduate nursing students, and 16 years as Executive Director and Administrator of Rush Home Care Network. She is currently associate chairperson for the Department of Community Health Nursing, and has authored several articles including "A Paradigm Shift for the Home Care Provider" in Home Care & Managed Care: Strategies for the Future, *"Home Care" in* Gerontologic Nursing, *and "Change and our Home Care Paradigms" in* Home Healthcare Nurse. *She received her B.S.N. and M.A. in nursing of children from the University of Iowa, and her D.N.Sc. in nursing from Rush University.*

From a Culture of Safety to a Culture of Excellence

Quality Science, Human Factors, and the Future of Healthcare Quality

Authors

Martin D. Merry, M.D., C.M., Consultant, Senior Medical Advisor to New Hampshire Hospital Association, and Associate Professor of Health Management, University of New Hampshire, Exeter, NH

Jeffrey P. Brown, M. Ed., Principal, System Safety Group, Peterborough, NH

Could this happen in your community?

Martin D. Merry—Consider the following case study from a front-page story of a local newpaper: Parents bring their nineteen-year-old daughter to a hospital emergency room. Findings include a rapid pulse, severely low blood pressure, and an abnormal electrocardiogram. After five hours, her condition stabilizes and she is discharged, with instructions to see her family physician the next day, which she does. The family physician sees her and orders an echocardiogram for the next day, assuring the family that she will call them with the results. The echo is done the next day. When they do not hear from the family physician, the parents assume that all is well. However, the echo is not actually seen by a physician until the next afternoon (a Saturday, three days after the ER visit). The cardiologist reads the echo as "severely abnormal" and recommends a follow-up. But he doesn't notify anyone; instead, he dictates a report (which finally reaches the family physician five days later). The day after the cardiologist dictates his report (Sunday), the mother finds her daughter dead in bed. The autopsy report indicates multiple pulmonary emboli (i.e., blood clots that have traveled to the lungs) as the cause of death.

The U.S. healthcare system's dark side is revealed

The National Academy of Science Institute of Medicine's (IOM) dramatic November 1999 release of *To Err Is Human* simultaneously signaled the end of medicine's 2000-year-old tradition as a virtually pure craft endeavor and announced the beginning of health care's concurrent industrial, information, and consumer revolutions. This IOM report made it starkly clear that the case study above is not an aberrant fluke that might be written off as an extremely rare tragedy in an otherwise exemplary healthcare system—arguably "the best in the world."

Rather, in a direct and effective challenge to this idyllic myth, *To Err Is Human* revealed that medical error causing injury, and even death, was a common, daily occurrence throughout the U.S. healthcare system. Americans were suddenly jolted into a stark new realization: A system remarkable for its technological achievements is inextricably linked with what we now realize as a parallel dark side. It is now public knowledge that modern health care is indeed miraculous in its technological capability, but also potentially dangerous—even lethal—in its execution.

The root cause of health care's current dilemma: Its craft model basis

To Err Is Human clarified that the root cause of such preventable tragedies as the case study above is rarely attributable purely to incompetent or careless caregivers. Indeed, all three physicians involved in this unfortunate young woman's care were fully qualified members on the medical staff of an excellent community hospital. While the individual decisions of any of these physicians might be subject to question, the IOM report makes clear that the deeper issues of causation in such cases are embedded in the environment in which physicians practice. The root cause of the vast majority of such tragic cases is this: The healthcare (non)system on which we rely in our most vulnerable moments has in recent decades simply grown (1) far too large and complex for the craft model on which it was built and (2) far more dangerous than anyone, until recently, has realized.

Health care's "sigma gap"

Some comparisons allow perspective on health care's current dilemma. Systemic-error or defect rates are expressed in terms of sigma units, with a high sigma value correlating with low defect rates (see Figure 1 below). World-class manufacturing competitiveness dictates that firms unable to generate defect rates in the five- to six-sigma range are likely to fail. In fact, the airlines are now achieving 0.43 deaths per million passengers, well above the near-perfection of six sigma for this key safety indicator. In stark contrast, healthcare measures generally fall into the two- to four-sigma range. For example, a recent *New England Journal of Medicine* article found that 2% to 8% of the patients who visit hospital emergency rooms with heart attacks are misdiagnosed as having something else. This percentage of missed diagnoses, not unlike the fatal example of our case study, translates to 20,000 to 80,000 errors per million heart attack patients who visit hospital ERs.

Figure 1. Sigma Speak.

Sigma Level	Defects per Million
1	690,000
2	308,000
3	66,800
4	6,210
5	230
6	3.4

It is important to withhold emotionally based judgment regarding the enormous "sigma gap" that currently exists between health care's current state and that which it needs to achieve in the future. As noted above, health care's current sigma state is directly attributable to its craft-model basis, which is now seriously inadequate for its complexity.

The underlying assumptions of health care's craft model

Because this craft model is so fundamental to health care's current dilemma, it warrants brief exploration. The craft model of healthcare system development was founded on a number of assumptions that were once arguably valid but no longer apply. These assumptions are as follows:

The underlying assumptions of health care's craft model, continued

1. Diagnosis and treatment of disease are complex tasks that require very specialized knowledge and experience.

2. In general, these tasks are beyond the knowledge of the vast majority of potential patients.

3. Physicians are the only professionals qualified to master the diagnosis and treatment processes.

4. Physicians acquire this mastery via a long period of book learning, hands-on experience, and mentoring by senior masters; hence, they are actually "master craftspersons."

5. Given the Hippocratic dictum of "First, do no harm," it is best to build health systems around these physicians and other healthcare professionals with confidence that "If it's good for the caregivers, it must be good for the patients."

Prior to the latter decades of the twentieth century, these assumptions appeared to well serve caregivers and patients alike. Given the relative simplicity of the medical encounter, and the relatively few effective treatments available during that earlier era, the potential problems inherent in these assumptions never came to light. But the exponentially expanding complexity of healthcare technology and delivery of recent decades has rendered these assumptions not only obsolete, but also dangerous.

What makes our current system so dangerous?

Why is this so? How can it be that a set of assumptions at the core of the world's most effective healthcare system can suddenly become obsolete, even dangerous? Because a highly complex, inherently high-risk system cannot be built on a craft model. Such complex systems require sophisticated design elements to prevent the multiplication of human error indigenous to highly labor-intensive, complicated endeavors such as health care. Most important, such systems must not be built around how individual caregivers—including physicians—practice; rather, they must be built around how those served by the system flow through it.

Why haven't healthcare professionals created safer systems? Again, the answer to this question can be traced back to the craft model and two fatal defects that only recently have come to light.

1. It relies on humans checking humans as its basic safety mechanism, thus actually compounding the likelihood of total *process* error in complex, multiple-caregiver settings.

2. It is essentially "blind" to the systemic complexities of modern health care—especially today's de facto high-risk environments that, via their complexity, actually increase the likelihood of human error.

"Human-error-proofing" systems are nonexistent

To mitigate the harm-generating effects of inevitable human error, complex high-risk environments require careful design and extensive testing. This concept is wholly incorporated into such high-risk endeavors as aviation, where autopilot systems do all the "heavy lifting" of flying a complex modern jet and thus preempt

"Human-error-proofing" systems are nonexistent, continued

the inevitable errors of human pilots and their potential adverse consequences.

In contrast, health care (with rare exceptions, such as anesthesiology) has never built into its core processes such "human-error-proofing" systems as the autopilot. Health care still relies almost totally on the knowledge and skills of caregivers—essentially denying the reality that no matter how knowledgeable, skilled, or "caring/careful" these people might be, they will still make errors at an irreducible minimum human rate—probably around four sigma. And some of these errors will generate harm—a certain percentage of which will be fatal to patients, as in the case study outlined earlier.

Managing safety in high-risk systems: Health care's craft tradition vs. those of other modern high-risk endeavors

Feed-forward vs. feedback control systems

Jeffrey P. Brown—As discussed by Dr. Merry, the release of *To Err Is Human* (IOM 1999) fundamentally altered our perspective on the reliability of healthcare systems in the United States. This report estimated that preventable medical error causes up to 98,000 deaths each year—a staggering number, and a public safety concern of the highest order. At its core, this untenable rate of preventable death is a function of the widening gap between health care's social and technical complexity and its pre-industrial managerial capability. Fortunately, other high-risk industries have pioneered approaches to proactively managing safety that can narrow this gap.

James T. Reason (1997) discussed administrative control in systems where outcome reliability is dependent on a blend of rule-based and knowledge-based problem solving, such as health care. He concluded that such systems require a balanced type of administrative control, which he characterized as "feed-forward" and "feedback." Feed-forward administrative control is an approach to managing organizational processes wherein decisions flow from the "top" to the "bottom" of the organization (see Figure 2 on the next page). Healthcare systems predominantly conform to a feed-forward administrative control model. The lack of familiarity of feedback administrative control (i.e., bottom-to-top checks and balances) in health care presents an experiential and structural barrier to the adoption of balanced, proactive system approaches to safety.

Active failures and latent failure conditions contribute to adverse events

Reason cites two ways in which decisions and actions contribute to adverse events in organizations. The most obvious is direct operational errors of action, inaction, or decision making ("active failures") occurring at the delivery level of the organization. Active failures may include violations, which constitute deviation from policy or procedure. Less obvious are "latent conditions" for failure. These are conditions that can provoke operational error under certain circumstances, or present hazards of their own accord (see Figure 3 on the next page).

Latent conditions for failure include inadequate training, unworkable procedures, and poor or inadequate technology (Reason 1997). Undue time pressure,

Feed-forward vs. feedback control systems, continued

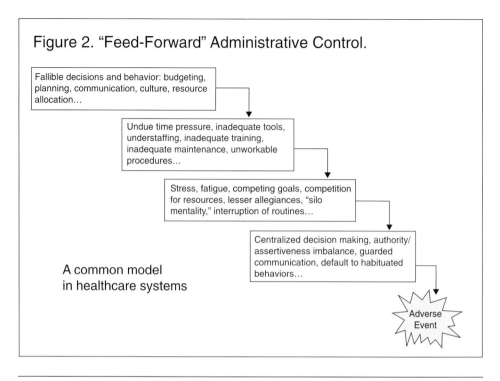

Figure 2. "Feed-Forward" Administrative Control.

Fallible decisions and behavior: budgeting, planning, communication, culture, resource allocation…

Undue time pressure, inadequate tools, understaffing, inadequate training, inadequate maintenance, unworkable procedures…

Stress, fatigue, competing goals, competition for resources, lesser allegiances, "silo mentality," interruption of routines…

Centralized decision making, authority/ assertiveness imbalance, guarded communication, default to habituated behaviors…

A common model in healthcare systems

Adverse Event

Active failures and latent failure conditions contribute to adverse events, continued

understaffing, and fatigue are also latent conditions for failure. Weick (1990) identified four processes that can emerge under stress that are an inherent source of vulnerability in human systems, including "interruption of important routines, regression to more habituated ways of responding, breakdown of coordinated action, and misunderstandings in speech-exchange systems." Fragmentation of effort in clinical environments is both a manifestation and a source of stress. High workloads, fatigue, emergent crises, time pressure, over-tasking, and understaffing are common

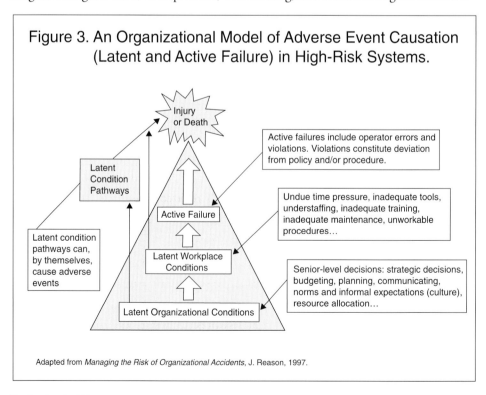

Figure 3. An Organizational Model of Adverse Event Causation (Latent and Active Failure) in High-Risk Systems.

Injury or Death

Latent Condition Pathways

Latent condition pathways can, by themselves, cause adverse events

Active Failure

Latent Workplace Conditions

Latent Organizational Conditions

Active failures include operator errors and violations. Violations constitute deviation from policy and/or procedure.

Undue time pressure, inadequate tools, understaffing, inadequate training, inadequate maintenance, unworkable procedures…

Senior-level decisions: strategic decisions, budgeting, planning, communicating, norms and informal expectations (culture), resource allocation…

Adapted from *Managing the Risk of Organizational Accidents*, J. Reason, 1997.

Active failures and latent failure conditions contribute to adverse events, continued

stressors acting on delivery-level personnel in the U.S. healthcare system (IOM 2001). Escalating political and financial pressures on the healthcare system are exacerbating these and other problems.

Latent conditions for failure frequently arise from high-level decisions and actions and are often deeply rooted in organizational culture (Carthey et al. 2000; Helmreich and Merritt 1998; Maurino et al. 1995; Schein 1996; Patankar and Taylor 1999). Examples of latent organizational conditions for failure include the strategic decisions and actions of the organization's managers (e.g., budgeting, planning, and resource allocation) and of government and industry regulatory entities. Latent failure conditions in the clinical space, such as undue time pressure, understaffing, inadequate tools, and managerial deficiences, are typically made visible in healthcare systems only after they precipitate unsafe acts and/or adverse outcomes. Feedback loops designed for the *proactive* identification and correction of latent failure conditions are virtually nonexistent in healthcare systems (IOM 1999, 2001).

The need for "error-tolerant" management systems

The term *error-tolerant* is not intended to convey that it is "okay" to have systems that provoke a high rate of error, injury, and death. In the parlance of human-factors science, the term is intended to convey the need to accept that errors will occur in any system, no matter how well managed, and that early identification and analysis of errors can provide an opportunity for the proactive correction of conditions that are unsafe.

As discussed above, research in high-risk domains has revealed that preventable errors are predominantly a consequence of workplace conditions that provoke error and that originate in organizational processes. Complex productive systems continually generate these *error-provoking* (i.e., latent failure) conditions, which often trigger a "near-miss" prior to inducing an injury or death. The occurrence of a near-miss indicates not only the presence of a new or recurrent threat to safety but also the opportunity to proactively identify and correct the root causes of the condition(s) that induced it. In addition, near-misses enable the identification and reinforcement of the behavioral and/or technological safeguards that might have prevented the near-miss from becoming an adverse event.

These are the principal reasons for the development of proactive risk management strategies designed to "flag" near-misses. Such management systems are commonly referred to as error-tolerant because they treat error and near-misses as inevitable and as a manifestation of system vulnerability, not individual deficiency or fault (Reason 1997).

Among the key requirements for establishing an error-tolerant management system are the means and methods for making error visible; that is, recognizing and reporting near-misses so that root causes can be identified and corrected. To manage the risk of active failure and to correct latent conditions, a methodology is required that both (1) actively prevents, traps, and mitigates error in clinical space (Helmreich 2000; Helmreich and Foushee 1993; Mudge 1998; Uhlig et al. 2001) and

The need for "error-tolerant" management systems, continued

(2) generates proactive feedback (Figure 4) on near-misses and unsafe conditions to enable root-cause analysis and correction of latent-failure states that have arisen as a function of fallible, "upstream" management decisions (Reason 1990, 1997).

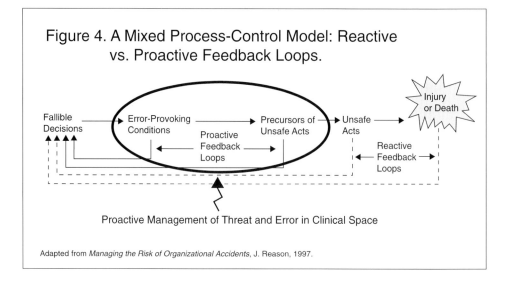

Figure 4. A Mixed Process-Control Model: Reactive vs. Proactive Feedback Loops.

Adapted from *Managing the Risk of Organizational Accidents*, J. Reason, 1997.

Managing the risk of active failure: Team-based approaches to actively limiting error in clinical space

Teams often perform better than individuals

There is considerable evidence that teams accomplish most safety-critical tasks better than individuals (Ginnett 1997). Multiple research findings reveal that groups perform better than individuals under stress (Suchman 1987; Weick 1990; Weick and Roberts 1993). Several authors have emphasized the need to redesign healthcare systems to support interdisciplinary teamwork from a safety and efficiency perspective (Nolan 2000; Reason 2000; Helmreich 2000; IOM 1999, 2001; Uhlig et al. 2001). Recent studies in aviation have outlined the fundamental elements of team-based approaches to safety.

Team-based methods for managing the potential for active failure in high-risk operational environments are commonly referred to as Error Management or Crew Resource Management (Helmreich 1996; Mudge 1998; Reason 1997). Klinect and Helmreich (2000) conducted an analysis of 4000 Line Observation Safety Audits (LOSAs) of airline crews throughout the world. Their findings identified the following key behavioral countermeasures and safeguards against active failure that are exercised by flight crews:

• Team building: leadership and communication environment.

• Planning: briefing, statement of plans, workload assignment, and contingency planning.

• Execution: monitor and cross-check, workload management, vigilance, and automation management.

• Review and modification: evaluation of plans, inquiry, and assertiveness.

Teams often perform better than individuals, continued

These countermeasures and safeguards are being increasingly incorporated, or further developed, as foundation elements in the error-management training programs of major airlines in the United States and other nations (Brown 2001; Gunther et al. 2001). These countermeasures and safeguards have also been adapted and piloted in health care with promising results (Uhlig et al. 2001).

The role of structured communication methodologies

The LOSA findings of Klinect and Helmreich (2000) are consistent with the findings of Taylor and Patankar (2000), who have characterized the behavioral countermeasures and safeguards against threat and error in terms of "structured communication," where *structured* refers to the requirement for personnel with interdependent roles to communicate, and *communication* refers to the required use of a standardized communication process for collective judgment and decision making (Mudge 1998; Weick 1995, 2001). Structured communication, as an approach to actively limiting error in high-risk operational settings, establishes the essential behaviors of collective practice, or *sensemaking*. Sensemaking, per Weick (1995), can be viewed as an approach to assessing a situation as a primary function of judgment and decision making under uncertain circumstances.

While structured communication methodologies are effective in limiting the risk of active failure at the delivery level of a system, they are not sufficient as a lone strategy to mitigate the risk of adverse events in complex organizations. Establishing a system for continual identification and correction of conditions that provoke error, or otherwise create a hazard, is critical to sustaining a proactive organizational approach to managing safety (Reason 1997; Vincent and Adams 1999; Wiegmann et al. 2000).

Managing latent failure conditions: Feedback loops and data classification needs for early warning and proactive correction of latent failure states

Applying procedures from the aviation industry to health care

In addition to providing teams with a means of avoiding, trapping, and mitigating error at the "delivery end" of high-risk systems, structured communication error-management methods can provide teams with practical approaches to providing early warning of latent failure conditions that have arisen through organizational processes. Teams can operate proactive safety-feedback organizational-learning loops by bringing error-provoking and otherwise unsafe conditions to light as a function of debriefing. In aviation, flight crew debriefing is ordinarily conducted at the conclusion of a flight (or mission). In health care, debriefing can be conducted at the conclusion of a shift, preferably at the overlap between shifts, to convey critical information to the team coming on duty.

To be useful for organizational approaches to safety, information garnered through debriefing must be described and classified in a manner that enables identification of performance factors beyond the local realm of individual, team,

Applying procedures from the aviation industry to health care, continued

task, and technical performance. Data that guides the identification and analysis of "upstream" organizational factors in near-misses or suboptimal/unsafe outcomes must be captured as well.

Approaches to classifying and analyzing human and organizational factors in error-provoking/unsafe conditions have been validated for incident analysis and safety intervention in aviation and are being studied in health care (Sarter 2000; Vincent and Adams 1999; Wiegmann et al. 2000). These methods of identifying, analyzing, and intervening in conditions that adversely influence human performance are aimed at identifying and correcting failed or absent defenses or safeguards against active failure (Reason 1990, 1997; Vincent and Adams 1999; Wiegmann et al. 2000). These failed or absent defenses and safeguards are, effectively, latent organizational and workplace conditions that induce active failure.

Five important perspectives on human error

The most effective of these investigation, classification, and analysis methods embrace five primary perspectives on human error, revealed in recent reviews of literature on human error and error analysis (Wiegmann et al. 2000, 2001). These perspectives, or models, and their key emphases, are as follows:

Cognitive. This perspective on error examines such factors as attention allocation, pattern recognition, and decision making. According to Wiegmann and Shappell (2001) and other researchers, cognitive frameworks—while useful in determining judgment, procedural, and response-execution errors—have not typically addressed contextual or task-related factors in error, individual physiological issues, or organizational factors (Rasmussen 1982; Wickens and Flach 1988).

Ergonomics and systems designs. These models examine the interdependencies of individuals, tools, machines, and the workplace, highlighting human/machine interface issues and anthropometric requirements of tasks. They do not provide in-depth examination of cognitive, social, and organizational factors (Heinrich et al. 1980; Wiegmann and Shappell 2001).

Medical. Medical models focus on an understanding of physiological factors in human performance. Under such models, the physiological condition of delivery-level personnel can be viewed as a "resident pathogen" (Reason 1997) in human/machine systems that, when triggered by workplace conditions, manifest as error. While these approaches generate important information on fatigue, illness, and other factors in human performance, they are most useful for identifying physiological factors as contributory issues in incidents or adverse events. These models have been useful in shaping policies on work scheduling and shift rotation (Lauber 1996; Weigmann and Shappell 2001).

Psychosocial. Psychosocial perspectives on error embrace work within complex productive systems as a social endeavor with multiple interactions among personnel with interdependent roles (Wiegmann and Shappell 2001). Human performance is viewed as being directly influenced by the nature or character of interactions among group members (Helmreich and Foushee 1993). The essential theme of

Five important perspectives on human error, continued

psychosocial models is that errors and adverse events occur when there is a breakdown in group dynamics and interpersonal communication (Weick 1990; Wiegmann and Shappell 2001). With few exceptions (Lynch 1996; Mudge 1998; Patankar and Taylor 1999; Taylor and Patankar 2000), these approaches have not generally identified "upstream" factors that bring about these "breakdown conditions" at the system operator level.

Organizational approaches. Organizational system models of error causation have been utilized in a range of industrial settings for many years and are being advocated and adapted for application to healthcare systems (Ammons et al. 1988; Heinrich et al. 1980; Reason 1997, 2000; Uhlig et al. 2001). These approaches consider adverse events to be the product of the unexpected confluence of latent and active failure states. They identify and analyze individual, team, task, technical, and organizational factors in near-misses and adverse events to develop intervention strategies for the correction of failed or absent defenses and safeguards.

Organizational models view front-line personnel as the last defense against a chain of fallible decisions' progressing through the organization to trigger an adverse event. These approaches are set apart from other models in that they place emphasis on the decision process at all levels of the organization (Reason 1997).

The quest for a safety culture

Culture and behavior can be thought of as "different sides of the same coin" (Uhlig et al. 2001). Just as it is reasonable to assert that organizational culture yields behavior, it is also reasonable to assert the reverse. New organizational behaviors that are embedded and accepted into the daily routines of practitioners can eventually define a new organizational culture (Schein 1996; Uhlig et al. 2001). The use of a structured communication methodology as an agent for the concurrent implementation of system error management and cultural change has been studied in corporate aviation settings and health care (Taylor and Patankar 2000; Uhlig et al. 2001). Such methodology actively limits error and, through debriefing, provides a means of prioritizing and analyzing information for proactive correction of error-provoking conditions in clinical space. These examples suggest that sustainable cultural transformation and continual monitoring of organizational-safety health can be achieved by careful redesign of interaction and communication at the delivery level of high-risk systems.

Despite the significant attention currently directed toward the development of reporting systems and "safety culture" in health care, very little is being written about practical, concurrent approaches to achieving error reduction and organizational change. A safety culture (or learning culture) is, fundamentally, characterized by collaborative interaction, an objective communication process, and nonpunitive approaches to learning from outcomes, whether favorable or adverse.

Establishing and sustaining a learning culture in healthcare systems require new communication behavior. The communication behaviors and social structure that enable active error reduction are built on forthright communication and an

The quest for a safety culture, continued

objective process for judgment and decision making as a team, and they embrace suboptimal outcomes as an opportunity to improve individual, team, and organizational performance.

These norms can be continually reinforced by a "mixed" administrative control model, wherein the organization requires team debriefing and feedback on system performance and then ensures rapid analysis and correction of the causal factors underlying problematic conditions reported by the team. Through prompt response, the organization continually reinforces the collaborative social structure and communication behavior necessary for active error reduction and correction of latent failure conditions.

According to Heinrich et al. (1980), the most robust methods of error and accident prevention "are analogous with the methods for the control of quality, cost, and quantity of production." The fundamental methods and principles for systemic improvement of decision making, coordinated action, and production outcomes, which have been studied for several decades by organizational psychologists, are comparable to the principles and methods for managing active and latent failure in high-risk operational settings. These methodologies provide for continual monitoring and improvement of behavioral and technological countermeasures and safeguards against active failure at the operator level of the system while providing feedback for correction of system factors implicated in analysis of near-miss data. To achieve a six-sigma safety standard in health care, cultural change is imperative.

Creating a new culture of safety and excellence

The present healthcare culture

Martin D. Merry—Health care cannot move beyond the approximately two- to four-sigma defect rate inherent in even the best of human systems without (a) addressing the powerful negative forces inherent in its present culture and (b) developing a leadership process that can build an entirely new "post-industrial" healthcare culture that fully incorporates the best of modern quality and error-prevention sciences.

It is probably erroneous to speak of a "healthcare culture" as if it were a monolithic entity. In fact, a basic reality of health care's craft-based culture is fragmentation. This fragmentation—which patients experience as enormous difficulty when moving from one element (e.g., a hospital) to another (e.g., a physician's office)—is a natural outgrowth of both the structural isolation of healthcare elements (hospitals, physicians' offices, long-term care facilities, visiting nurse services, and so forth) and the professional isolation of the various categories of caregivers of present-day health care. For instance, consider how physicians relate to hospitals in the typical community hospital setting. Figure 5 on the next page represents a generic hospital organizational structure. Of note, the "medical staff" (i.e., the physicians) is actually a distinct entity, responsible primarily to its own elected leadership, and not under the purview or authority of executive management. As any

The present healthcare culture, continued

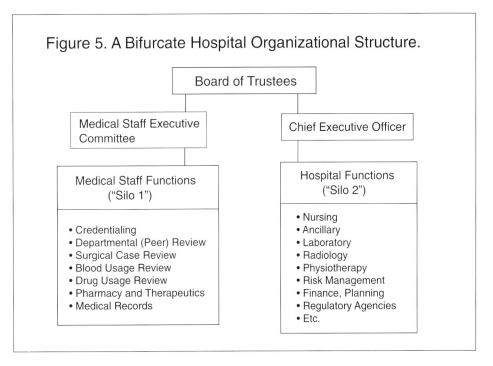

Figure 5. A Bifurcate Hospital Organizational Structure.

student of management might predict, such a structure breeds isolation of two essential components of the hospital—and, with its ambiguous authority, inevitable power struggles between physicians and institutional management.

This isolation and fragmentation were plainly evident in the case study noted earlier. All three physicians who played roles in this tragedy were practicing in complete isolation. There was no evidence that they had any communication among themselves regarding this fatally ill young woman, and they most certainly had no common medical record that might have united them in their caring for her. Is it credible that the cardiologist, for example, would not have called someone if he had been aware of how this young woman appeared in the ER three days earlier?

While the newspaper story didn't spell this out directly, those who are familiar with the current fragmentation of health services know that this cardiologist likely had access to no more information than the echocardiogram printout that he read and dictated. Had he been aware of the total clinical picture of this dying young woman and had made that telephone call, this case might instead have been a near-miss.

As the airline industry has clearly taught us, such near-misses are extraordinarily valuable learning opportunities. In this case, physicians and hospital personnel alike might have learned much from such a near-miss regarding the potentially dangerous lack of continuity from hospital ER to private physician's office to hospital cardiology lab, and so on.

In addition to this structural impediment to unified action, consider the different professional profiles of physicians and healthcare managers (see Figure 6 on the next page). It is clear from this figure that the two separate and distinct training traditions of physicians (M.D., D.O.) and of healthcare administrators (M.H.A.,

The present healthcare culture, continued

Figure 6. Health Care's Professional Cultures.

Physicians	**Managers**
• Doers	• Designers
• 1:1 Problem Solving	• 1:*n* Problem Solving
• Reactive	• Proactive; Long Time Span before Results
• Immediate Response	• Long-Term Response
• Deciders	• Delegators
• Autonomous	• Collaborators
• Independent	• Participative
• Patient Advocate	• Organization Advocate
• Professional Identity	• Organizational Identity
• Independent Professional	• Interdependent Professional

From "Can Doctors and Administrators Work Together?" by Sandra L. Gill (*Physician Executive*, Sept./Oct. 1987).

M.B.A.) generate professional behavioral profiles that are very likely to misunderstand one another, and thus be in conflict.

A "culture of blame"

As if these impediments to a healthcare culture of collaboration around creating safe, high-performing systems weren't enough, the IOM correctly identified health care's "culture of blame" as a third major barrier to safer healthcare systems. What is a culture of blame? It is essentially one of fear and finger-pointing, based on the expectation that if one is found out to be in error, punishment is likely to ensue. If one is under duress in such a culture, the natural human protective reaction in such a culture is to defend oneself, to find somewhere else to assign blame.

And how does health care develop this culture? Yet again, we encounter its craft tradition. As a result of their training, craftspersons introject the notion that "Because I'm so highly trained and no one else can do what I do, I'm totally responsible for all the consequences of my actions." In medical training, this translates to a tradition of developing a great sense of responsibility for all of one's actions: "I check and double-check, because if I make an error, someone may die."

While this statement is literally true and has an obvious benefit in creating a sense of personal responsibility, it also inadvertently places an impossible burden of perfectionism on the apprentice in training. This is an unintended consequence in a high-risk environment—one nearly totally dependent on human checking systems. It creates a constant fear of making a mistake (exacerbated by a harsh medico-legal climate) and a consequent adoption by those "implicated" of various psychological and legal "defenses," such as denial, rationalization, and blaming others.

In fact, even as of this writing this "culture of blame" staggers on like a soulless

A "culture of blame,"
continued

zombie. Our case study's final ending illustrates this point: The implicated hospital and physicians settled the ensuing malpractice lawsuit for a huge sum, and the ER physician was publicly censured by his state medical board.

Does anyone believe that these measures will create a safer environment for patients at this hospital? Is this ER physician likely to become a "better physician" as a result of his front-page, public humiliation by the state board? Is the cardiologist any more likely to pick up the telephone after he has read another "severely abnormal" echocardiogram as a result of this experience? Will the hospital establish a task force to address the serious danger of the fragmentation of its services as a result of its huge malpractice settlement? Will the hospital's CEO, board of trustees chairperson, and/or president of the medical staff meet with members of the community to share with them what they are doing to create safer care in their hospital?

An honest answer to all these questions is "perhaps," but as of this writing there is no documented evidence that fear of litigation or punishment has ever improved practice at either the practitioner or the institutional levels.

Getting from here to
where we need to be

Health care desperately needs creative, robust solutions to its safety crisis. The IOM calls for nothing less than a new healthcare system for the twenty-first century. As noted earlier, health care's 2000-year-old craft-based culture is not one that welcomes the innovation now essential to traversing its industrial/information/consumer revolutions. Figure 7 illustrates the contrast between what "has been" and what "must become." Leadership, management science (including such elements as Six Sigma management and human factors), and conscious development of organizational culture—which were all relatively unimportant

Figure 7. Craft- vs. Systems-Based Health Care.

Feature	Craft-Based	Systems-Based
Core Focus	React to One Patient at a Time	Plan for 1+ Population
Leadership	Largely Irrelevant	Key Success Factor
Management Science	Minimally Relevant	Key Success Factor
Organizational Culture	Largely Irrelevant	Key Success Factor
Key Quality Factor(s)	Individual	Individual + System
Quality Capability	Two to Four Sigma	Five to Six Sigma

in the traditional craft-based system—are now vital to health care's success.

But health care's transformation is inevitable. One definition of insanity is continuing to do the same things you've always done in the past and expecting a different result. We will continue to get the same two- to four-sigma results until we build safe infrastructures. Public leaders and healthcare leaders alike must grasp the

Getting from here to where we need to be,
continued

fundamental truth that modern medical miracles can occur and progress only in a scientifically sophisticated—and inevitably complex and high-risk—system. All must accept that the price of continued clinical innovation will be the investment in safe infrastructures and the creation of latent workplace conditions that, as the IOM has suggested, "make it hard for people to err, and easy to do what's correct." In fact, our current two- to four-sigma (non)system is enormously costly in terms of waste, rework, inefficient use of professional time, and human suffering of patients and caregivers alike.

Is Six Sigma a realistic goal?

Can health care realistically pursue Six Sigma? In terms of vital measures of patient safety, we can, like the airlines—and indeed must. In fact, anesthesiology, as a result of a sustained effort and the importation of high-risk system-design concepts, is already approaching Six Sigma in anesthesia-related mortality. (And it is yet another example of health care's fragmented culture that the knowledge gained in this effort has not diffused into other medical areas.) But, like the anesthesiologists, for a time we will need to import help from quality-management and human-factors specialists outside the healthcare field. The simple fact is that neither clinicians nor current health administrators are being trained in systems thinking for clinical design.

Healthcare leaders will need conceptual models on which to build future systems. At least five key elements will characterize those leaders who successfully transcend today's healthcare cottage industry in their pursuit of Six Sigma excellence. These healthcare leaders will master the following:

- A much-improved understanding of the population's health needs and how people actually "flow" through various care subsystems.
- Systems thinking and quality-related sciences.
- Unprecedented collaboration between clinical and management science.
- The creation of a new "medico-management" culture—one that maintains the best of both clinical and management traditions, even as it casts off counterproductive elements from both.
- Abandonment of hierarchical, "command-and-control" leadership models in favor of the collaborative, dispersed team-based leadership models that always work better in complex, high-risk environments.

The necessary tools are available

For those who can begin working from such conceptual frameworks, the tools are already available. In fact, health care has already experienced the "total quality management" (TQM) wave that occurred between 1988 and its relative demise around 1995. But today there is considerable residual knowledge embodied by such organizations as Boston's Institute for Healthcare Improvement and Salt Lake City's Intermountain Health System. In essence, future healthcare design will revolve around TQM's process improvement concepts, which are now augmented by modern Six Sigma and human-factors science.

The necessary tools are available, continued

Because it has no other real choice, health care will embrace management systems guru W. Edwards Deming's dictum that improving quality through process improvement simultaneously decreases cost and enhances value delivered. This is precisely the formula needed for an industry grotesquely bloated from the combined cost of waste, inefficiency, and patient injury. With health insurance premiums spiraling out of control, and with no easy fixes on the horizon, could there be a better time for healthcare leaders to actualize Dr. Deming's dictum?

In fact, the transformation has already begun. One of health care's most gifted futurists, Jeff Goldsmith, created in 1995 the basic model from which Figure 8 is adapted. Though still well-entrenched in Stage 1 as of 1995, health care is obviously

---

Figure 8. The Journey: Health Systems (R)evolution.

Stage 1: Event-Driven	Stage 2: Value (Quality/Cost)	Stage 3: Healthy Community
• Payers seize power • Hospital usage decreases • Compete on price: slash/burn cost cutting • Primary care networks	• Control resource intensity • Redesign system: patient interface, human-factors applications • Total quality: clinical paths, disease management, outcome monitoring, etc. • Evolved physician networks	• Population-based • Health-status appraisal • Prevention emphasis with focused intervention • Case/disease management
The end of craft-based health care	**A bridge; the "edge of chaos"**	**Emergent, systems-based health care**

Adapted from Jeff Goldsmith, Ph.D.

---

now on the threshold of Stage 2 and on its way, via the efforts of true innovators and early adopters, to Stage 3 (elements of which already exist sporadically throughout the United States). Stage 2 is enormously promising, but it predictably will be a stormy sea to traverse as health care's 2000-year-old resident culture stages its last battle to preserve outmoded thinking and practices.

Is it revolution, evolution, or both?

As Charles Darwin said, "It's not the strongest of the species that survives, nor the most intelligent, but the one most responsive to change." This truly is a(n) (r)evolutionary time for health care. Its ancient culture is giving way, finally, to an adaptive change that the authors of this paper believe will give rise to health care far better than anything we have ever experienced to this time. We believe that the factors outlined herein hold promise of successfully addressing the "golden quintet":

- Significantly improved clinical outcomes

- Far better patient/customer satisfaction

- More humane, rewarding working conditions for caregivers

Is it revolution, evolution, or both?
continued

- Greatly improved safety

- Surprising cost constraint

In sum, we have the ability to create something beyond our fondest imagination. And the challenge to create it is not just for our healthcare leaders. Health care belongs to society as a whole, and we all have potential roles in the unfolding of its future. As we contemplate our individual roles in health care's transformation, each of us might ask these simple questions: "If not now, when?" "If not me/us, who?"

References

Ammons, J. C., T. Govindaraj, and E. M. Mitchell. 1988. "A Supervisory Control Paradigm for Real-Time Control of Flexible Manufacturing Systems." *Annals of Operations Research*, 15: 313–335.

Brown, Jeffrey P. Discussion with training captains from major world airlines. First annual Captains' Leadership Symposium, hosted by Delta Air Lines, Atlanta, Georgia, March 2001.

Carthey, J., M. de Leval, and J. Reason. 2000. "Adverse Events in Cardiac Surgery: The Role Played by Human and Organizational Factors." Pre-publication manuscript.

Ginnett, Robert C. "Building a Culture for Team Safety: By Design and by Default." From proceedings of the NTSB Symposium on Corporate Culture and Transportation Safety, April 1997.

Gunther, Don, Robert Helmreich, and Bryan Sexton. Continental Airlines CRM Program. Presentation at the eleventh International Symposium on Aviation Psychology, Columbus, OH, March 2001.

Heinrich, H., D. Petersen, and N. Roos. 1980. *Industrial Accident Prevention: A Safety Management Approach,* 5th ed. New York: McGraw-Hill.

Helmreich, Robert L. 1996. "The Evolution of Crew Resource Management." IATA Human Factors Seminar, Warsaw, Poland.

———. 2000. "On error management: lessons learned from aviation." *BMJ* 2000; 320: 781–785 (18 March).

Helmreich, R., and H. Foushee. 1993. "Why Crew Resource Management? Empirical and Theoretical Bases of Human Factors Training in Aviation." In *Cockpit Resource Management,* eds. E. Wiener, B. Kanki, and R. Helmreich (San Diego, CA: Academic Press), pp. 3–45.

Helmreich, R., and A. Merritt. 1998. *Culture at Work in Aviation and Medicine.* Brookfield, VT: Ashgate Publishing, Ltd.

Hollnagel, E. 1998. *Cognitive Reliability and Error Analysis Method.* New York: Elsevier Science.

ICAO. 1993. "Human Factors, Management and Organization." Circular 247-AN/148, *Human Factors Digest* no. 10. Montreal, Canada: The International Civil Aviation Organization.

IOM. 1999. *To Err Is Human: Building a Safer Health System.* Linda T. Kohn, Janet M. Corrigan, and Molla S. Donaldson, eds. Institute of Medicine, National Academy Press, pp. 17–25; 86–108.

———. 2001. Appendix B in *Crossing the Quality Chasm: A New Health System for the 21ˢᵗ Century.* Institute of Medicine, National Academy Press, pp. 322–335.

Klein, Gary, and David Klinger. "Naturalistic Decision-Making." Crew System Ergonomics Information Analysis Center (CSERIAC) Gateway, vol. II, no.1, Winter 1991.

Klein, G. 1993. "Naturalistic Decision Making: Implications for Design." CSERIAC SOAR 993 (1).

Klinect, James. 2000. "CRM: Lessons Learned from LOSA." From Bob Helmreich's LOSA presentation (on-line slide show). University of Texas human-factors research project, http://www.psy.utexas.edu./psy/helmreich/nasaut.htm.

Lauber, J. Foreword to *Basic Flight Physiology,* 2d ed., by R. Reinhart. New York: McGraw-Hill, 1996.

Lynch, Kevin. 1996. "Management Systems: A Positive, Practical Method of Cockpit Resource Management." In proceedings of the forty-first Corporate Aviation Safety Seminar. Orlando, Florida: The Flight Safety Foundation, pp. 244–254.

Maurino, D., J. Reason, N. Johnson, and R. Lee. 1995. *Beyond Aviation: Human Factors.* Brookfield, VT: Ashgate Publishing.

Mudge, Gordon W. "Airline Safety: Can We Break the Old CRM Paradigm?" *Transportation Law Journal* 25(2): 231–243 (Spring 1998).

Nolan, Thomas W. 2000. "System Changes to Improve Patient Safety." *BMJ* 320: 771–771 (18 March).

Patankar, Manoj S., and James C. Taylor. 1999. "Corporate Aviation on the Leading Edge: Systemic Implementation of Macro-Human Factors in Aviation Maintenance." SAE Technical Paper 1999-01-1596.

References, continued

Rasmussen, J. 1982. "Human Errors: A Taxonomy for Describing Human Malfunction in Industrial Installations." *Journal of Occupational Accidents* 4: 311–333.

Reason, James T. 1990. *Human Error.* New York: Cambridge University Press.

———. 1997. *Managing the Risk of Organizational Accidents.* Brookfield, VT: Ashgate Publishing.

———. 2000. "Human error: models and management." *BMJ* 320: 768–770.

Sarter, Nadine B. 2000. "Error Types and Related Detection Mechanisms in the Aviation Domain: An Analysis of Aviation Safety Reporting System Incident Reports." *International Journal of Aviation Psychology,* 10(2), 189–206.

Schein, Edgar. "Organizational Learning: What Is New?" Society of Organizational Learning, July 1996, http://www.sol-ne.org/res/wp/10012.html.

Senders, J., and N. Moray. 1991. *Human Error: Cause, Prediction and Reduction.* Hillsdale, NJ: Lawrence Erlbaum Associates, Inc.

Suchman, L. 1987. *Plans and Situated Actions.* New York: Cambridge University Press.

Taylor, James C., and Manoj S. Patankar. 2000. *The Role of Communication in the Reduction of Human Error.* Pre-publication copy provided by the authors.

Uhlig, Paul, et al. "Improving Patient Care by the Application of Theory and Practice from the Aviation Safety Community." Presented at the eleventh annual International Symposium on Aviation Psychology, Columbus, Ohio, March 6, 2001.

Vincent, Charles, and Sally Adams. 1999. *A Protocol for the Analysis of Clinical Incidents.* Association of Litigation and Risk Management, The Royal Society of Medicine, London.

von Thaden, Terry L. 2000. "Beyond CRM: Information Infrastructure in the Context of Distributed Practice among Flight Crews." Unpublished paper. Aviation Research Laboratory, University of Illinois at Urbana–Champaign.

Weick, Karl E. 1990. "The Vulnerable System: An Analysis of the Tenerife Air Disaster." *Journal of Management* 16 (3), 571–593.

———. 1995. *Sensemaking in Organizations.* Malden, MA: Sage Publications.

———. 2001. *Making Sense of the Organization.* Thousand Oaks, CA: Blackwell Publishers.

Weick, K., and K. Roberts. 1993. "Collective Mind in Organizations: Heedful Interrelating on Flight Decks." *Administrative Science Quarterly,* vol. 38(3).

Wickens, C., and J. Flach. 1988. "Information Processing." In *Human Factors in Aviation*, eds. E. Wiener and D. Nagel (San Diego, CA: Academic Press), pp. 111–115.

Wiegmann, D. A., A. M. Rich, and Scott A. Shappell. 2000. "Human Error and Accident Causation Theories, Frameworks and Analytical Techniques: An Annotated Bibliography." Technical Report ARL-00-12/FAA-00-7, FAA Civil Aeromedical Institute.

Wiegmann, D. A., and Scott A. Shappell. 2001. "Human Error Perspectives in Aviation." *International Journal of Aviation Psychology,* pre-publication copy.

Author information

Dr. Martin D. Merry received his undergraduate degree in industrial and labor relations from Cornell University and earned his medical degree at McGill University. He practiced general internal medicine for eight years while developing the role of Medical Director for Quality at St. Joseph's Hospital in Elmira, New York.

In 1981, Dr. Merry began a career devoted to consultation and education in the areas of quality, medical staff leadership, and organizational transition. He also teaches the quality management course in the Masters of Healthcare Administration program at the University of New Hampshire and serves as Senior Advisor for Medical Affairs for the New Hampshire Hospital Association.

Jeffrey P. Brown is Principal of System Safety Group. He specializes in the design and implementation of organizational approaches to safety, utilizing error-management methodology developed for systemic implementation in high-risk domains.

Editorial assistance for this article was provided by Cathy Kingery.

Process Management and Systems Thinking for Patient Safety

Authors

Joanne E. Turnbull, Ph.D., Executive Director, National Patient Safety Foundation, Chicago, Illinois

Introduction

To Err Is Human, a 1999 report from the Institute of Medicine, broke the news about the problem of medical error in health care. It said that two studies conducted a decade apart reported that our healthcare system had 44,000 and 98,000 deaths, respectively, caused by medical error per year. If you were to project those numbers out to today, it's probably even higher.

Process management is an incredibly important approach that could help eliminate medical errors and make patient safety a reality, but it is not being used consistently in health care right now. Many people have either experienced a medical error or know someone who has. In a four-month span in early 2001, I personally experienced four medical errors; none of them life threatening, and all of them related to broken or missing work processes. Medical errors occur because we do not have the work processes in place to support the many wonderful advances in medical technology.

But is process management enough? The answer is no—the problem is far too big. Physicians and nurses can run a pristine practice with process management in place, but unless the back-end factors—regulations, the legal system, and cultural factors—are fixed, it's all for naught. We must move away from a culture of blaming the person at "the sharp end"—in safety science, this term refers to the doctors, nurses, and providers—and fix the system.

The monetary cost of errors

Research shows just how costly these errors are. One study showed that two out of every 100 admissions experienced a preventable adverse drug event (ADE), resulting in an average increased cost of $4,700 per admission. Another study showed that the total national costs are between $17 billion and $29 billion, with direct healthcare costs representing over one-half of that amount. That wasted money would pay for many nursing positions and a great deal of uninsured care. The issue is not whether we have sufficient resources; rather, the issue is one of misallocation and misuse of resources. Employers are finally waking up to the fact that they are paying for error, and so they are beginning to hold the healthcare system accountable.

Tightly coupled work processes

In his book *Normal Accidents: Living with High Risk Technology*, Charles Perrow says that certain types of organizations are set up to produce accidents in the normal course of business, due to "tightly coupled work processes." These tightly coupled work processes can be represented by interlocking the fingers of two hands. And

Tightly coupled work processes, continued

although Perrow does not examine health care directly, healthcare organizations include many tightly coupled work processes just like the industries he examines. For example, if a physician writes an order that reaches a patient successfully, it has to do with many factors beyond how well the physician put pen to paper. Specifically, a successfully executed order relies on other professionals that the physician may not know personally, and more to the point, the physician is not likely to know the scope of their practice.

It is frightening to not be in control of work processes for which you are held accountable. And if a root cause analysis were conducted on the overarching causes of error in health care, I am certain that one of the biggest problems would be the disconnection between workers and their work processes.

Errors and missing processes

When a patient hears that a medical test must be repeated, it often means the test was lost, and a survey of outpatient physicians reported that lost diagnostic tests are the biggest cause of error. Why? Because there is no system in place to show that the test results were ever received, so important diagnoses are not made. Patients are told, "We will call you if something is wrong." But that assumes the test results were received in the first place, or that there is a process to determine if the test results were received or not. Currently that process is not there in most practices.

A medical error changes our lives

In 1995, a baby died at Hermann Hospital in Houston, Texas, where I was working. The resident physician on the case, awaiting her own diagnosis for lupus, misplaced a decimal point while recording the dosage of medication for the boy.

The way we approached the problem by a systemic focus on processes led to an article in the *NY Times Magazine* that was one of the first intelligent pieces written on medical error. It was seen as a turning point in the patient safety movement due to its focus on the system. The headline read, "Who's to Blame? It's the Wrong Question." The truth is that human beings make mistakes. The problem was that every process in place at Hermann Hospital let this mistake go through.

Understanding that our processes were not working

The death of this little boy was more than a serious event. It created a burning platform, and it brought us to our knees. We had to correct our way of working and analyze our processes immediately to ensure that it would not happen again. Clearly, our existing perceptions and methods were not working and needed to be corrected. To that end, we decided that (1) we needed to understand what the variations in our practice were and to determine what that actually meant. (2) We had to be educated in a new way. (3) We had to minimize the steps in the care process. And finally, (4) we had to work to abolish the culture of blame, punishment, and fear, the most difficult step of all.

Offsetting safety interventions

By building on these ideas, we were able to reduce serious medication errors by 50% (Figure 1 on the next page). But just correcting our processes was not enough

Offsetting safety interventions, continued

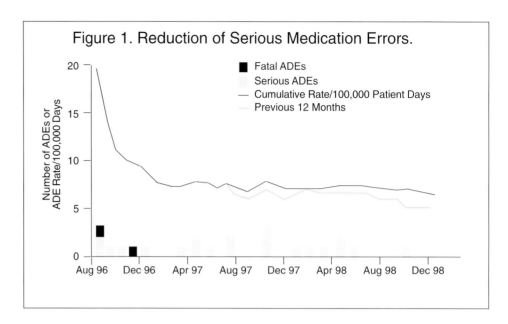

Figure 1. Reduction of Serious Medication Errors.

to eliminate errors. Let me give you an example. In the winter of 1998, every hospital in Houston, Texas, went on drive-by status, due to an outbreak of the flu. Gurneys were backing up as patients waited to be seen, and our census, which usually operated around 375, went up to 525. The next day, our chief operating officer said, "This is great. We managed an increase in census with no increase in staff." But further analysis showed that there were three serious medication errors during this time period. Now I bet an economic analysis would show that those three errors cost a lot more than the compensating staff would have cost.

This anecdote exemplifies a concept put forth in *Normal Accidents*: every time there is a safety intervention, there is a production pressure increase to add to it that negates the safety issue. Another example to illustrate this point comes from the airline industry. A colleague who studies that industry said that every individual flight is now safer, but overall, the whole system is more dangerous because more flights are added to offset the cost of safety interventions. We are running out of air bands in the sky. The same is true in health care. The population is getting older and sicker, but there is no additional money to support an overtaxed system.

The old look in patient safety

At the National Patient Safety Foundation (NPSF), we work to change perceptions about errors and patient safety. Our goal is to shift from an old look that blames people, to a new look that focuses on systems.

The old look in patient safety said (1) clinicians are supposed to be infallible, (2) bad mistakes happen only when people make mistakes, (3) people in organizations that fail are bad, and (4) blame and punishment sufficiently motivate carefulness. It's euphemistically called "the train and blame game."

The old look in patient safety did not prevent errors. In 1995, the Joint Commission on the Accreditation of Healthcare Organizations (JCAHO) experienced its own sentinel event. Several organizations that had recently been accredited with commendation had horrible things happen in the following weeks. For ex-

The old look in patient safety, continued

ample, a hospital in Florida amputated the wrong leg, and at the Dana-Farber Cancer Institute, a patient died from an overdose while undergoing chemotherapy. Obviously, accreditation based on the train and blame game was not good enough.

The new look in patient safety

The new look in patient safety says that (1) the risk of failure is inherent in complex systems. In other words, problems in health care are unavoidable; the denial is over. (2) Risk is always emerging. For example, we had a death in our hospital because CVVHD, a new kind of dialysis, was introduced. This procedure involves both pulmonary and renal function but there was no process to dictate which medical specialty should oversee the dialysis. Nobody was in charge. It was a risk that emerged that no one predicted. (3) Not all risk is foreseeable. (4) People are fallible, no matter how hard they try. Systems are also fallible. And (5) alert and well-trained clinicians are crucial.

Getting rid of blame and punishment may not happen in the current generation of healthcare workers, but it is the cultural change that must happen to make errors transparent and to learn from them.

High reliability organizations

Health care is a high-risk industry. Errors are everywhere, if you look for them. What we need to do is incorporate the lessons learned from other disciplines and industries, and adapt them to health care. Specifically, we can learn a lot from "high reliability organizations." High reliability organizations (HROs) admit that they have errors and discuss errors openly. The lessons they've learned show that you cannot eliminate error but you can mitigate it and control the risk involved.

High reliability organizations understand that (1) workers face increased process complexity that is often unmanageable. (2) Everybody is on information overload. For example, consider how many PIN numbers you need to keep track of. Now imagine what it is like to have thirty patients on an eight-hour shift, with insufficient resources and no supplies at hand. (3) Everyone—professionals and patients alike—has increased expectations for perfect outcomes. This is perhaps an unintended consequence of outcomes management. For health care, the situation is even more complicated because the system is taxed by new patient vulnerabilities. Patients are living longer, with multiple, serious chronic conditions.

Acknowledging and auditing risk

High reliability organizations acknowledge that risk exists. Three things are needed to change behavior: knowledge, motivation, and skill. Knowledge has two arms: (1) information or content knowledge, and (2) confronting denial. The hardest piece of all is overcoming that denial to acknowledge a problem exists.

High reliability organizations also audit risk. Auditing risk points out the biggest issue we are dealing with in health care today. The purpose of auditing risk is to learn from errors and near-misses. But to learn from errors, you have to count them. The biggest fear in health care today is that counting and learning from those errors exposes the organization and the individual to litigation. All of the work that we did

Acknowledging and auditing risk, continued

at Hermann Hospital to reduce medical errors by 50% is gone because after a merger, the new CEO was terrified that error reduction work would expose us to litigation.

The National Patient Safety Foundation is currently trying to engage trial lawyers' associations in a conversation on auditing risk but we have so far met with a great deal of resistance.

Process control

High reliability organizations depend a lot on process control. Process control involves:

- Rules and procedures,
- Training,
- Strategic redundancies,
- Teamwork development, and
- Mitigating decision making.

High reliability organizations have process documents that contain their *rules and procedures*; these organizations should be able to run regardless of the individuals involved, because all of the steps in their processes are documented. *Training* means training for performance, not training for knowledge. *Strategic redundancies* means mapping a process, admitting that there is the potential for error, and building safeguards into the system to prevent the error from occurring. *Teamwork development* is probably the biggest thing that has to happen for us in health care. And finally, *decisions* must be made by frontline workers, rather than by people who are not directly involved with what is going on, because the people at the front line understand their work processes and can make better decisions about the care involved.

Leadership buy-in and appropriate rewards

High reliability organizations also stress the need for leadership involvement. If there is no leadership buy-in at the organizational level, you may as well forget about making changes. Without leadership buy-in, the changes will not be sustained and the overall culture will not change.

Appropriate rewards are also very important to high reliability organizations. People are thanked rather than punished for reporting errors that occur and pointing out vulnerabilities that exist in the system.

The Onion Model

As I said before, process management isn't enough to solve all of the problems we are facing today. Figure 2 (on the next page) shows a socialization model of culture change called "the Onion Model." Change has to happen at every level of this model, and every component or agent in this system has to be aligned for patient safety to become a reality.

The levels of the Onion Model

The first level of the Onion Model is the practice level. The practice level must consist of not only the physicians, nurses, and pharmacists; it must include patients and families as well. The next level, the organizational level (which includes all

The Onion Model,
continued

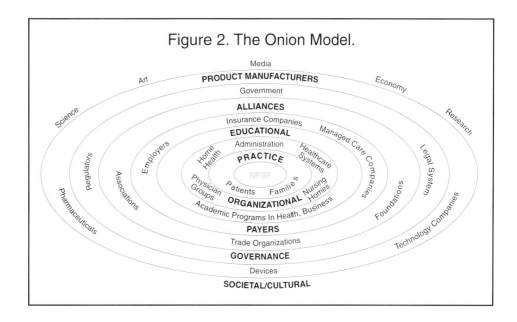

Figure 2. The Onion Model.

The levels of the Onion Model, continued

settings from home health to office to acute care to long-term care) must also undergo massive change to become high reliability organizations. At level 3, the educational level, all healthcare training programs—from consumer to practitioner to administrator—need a whole new way of thinking about education that includes the components of safety science over and above content expertise. Level 4 is the payer level, and this includes employers, insurance companies, and managed care organizations. Here, we are beginning to see incentives put in place. Alliances, associations, trade organizations, and foundations, which make up level 5, are now beginning to focus on patient safety. Level 6, the governance level, includes boards of directors of healthcare organizations, regulatory and accrediting bodies, and the legal system. Level 7 includes product manufacturers, including device makers, pharmaceutical companies, and technology companies. Level 8 includes societal or cultural influences, such as the media, the economy, and art.

Some examples of changes that are happening or need to happen are provided below.

Level 1: The Practice Level

The patient and family advisory council

Let's talk about our core business: what can be done for patients and families? In 1999 the National Patient Safety Foundation did not talk to families—we were scared of them. Calls that came in to the foundation from people who had lost loved ones to medical error were filed under "public complaints." Worse than that, we referred them to state medical societies, who are not prepared to deal with grieving people who are angry at an unresponsive system. So in May 2000, we took a big step and listened to what people had to say. In April 2001, we held our first patient and family advisory counsel meeting, which is now a formal part of the foundation to advise us on our policies.

This first meeting told us the many things that consumers need and want. (1)

The patient and family advisory council, continued

They want communication tools for different stages of care. Members of the council mapped out the different stages of care, the activities that occurred in those stages, and the communication tools that were needed in each stage. (2) They want a more open, less adversarial culture. (3) They want to be empowered to ask questions. (4) They want to change the expectation of consumers that perfect care is a myth. (5) They want on-site advocacy twenty-four hours a day, seven days a week, to prevent errors. (Most people do not know that hospitals hire patient representatives and patient advocates.) (6) They want to raise awareness among consumers, and (7) they want follow-up support when an error does occur. Follow-up support is important because the literature shows that if people receive compassionate, honest, adequate care following an error, they are less likely to sue. The reason they sue is to get an answer to what happened.

This type of interaction is important because it provides information to our foundation and helps us focus on what the public wants to ensure patient safety.

Level 2: The Organizational Level

Changing the organizational environment

Over and above the lessons from HROs, patient safety must be incorporated into the business plan, voluntary error-reporting systems must be implemented, and engineering concepts taken from other high-risk industries must be adapted and implemented into the healthcare setting.

The authority gradient

Teamwork is a key component of HROs. For effective teams to become a reality in health care, the authority gradient that exists in healthcare organizations must be eliminated. The authority gradient refers to interpersonal dynamics in situations of real or perceived power. In these situations, the truth is withheld if it is bad news. A story is told of a general who was promoted in the military and was told, "Congratulations, General. No one will ever tell you the truth again." Authority gradients exist as a hierarchy between a physician and a nurse, or an administrator and an employee.

To confront its own authority gradient, the airline industry trained pilots and copilots to communicate more effectively with each other. For example, copilots used to timidly ask the pilot, "Are you sure you want to take off?" Today, after crew resource management training, they feel empowered to say, "There is ice on the wings. You're not going to take off." As one copilot said, "Now I sit down with my pilot and say, 'I like you. You're a nice guy. I respect you, but we are in a high-risk environment. It's my job to watch you.'" This type of vigilance is exactly what we need in health care.

To confront the authority gradient at Hermann Hospital, we gave an Authority Gradient Card to the nurses that says, "We are a team. I have something to say." This is important because nurses have traditionally not felt empowered to ask questions. In the case of the baby's death, a nurse gave medication to the patient knowing that it was wrong, but was afraid to question the physician's decision.

The Iceberg of Ignorance

The Iceberg of Ignorance (Figure 3) is a model based on the work of a Total

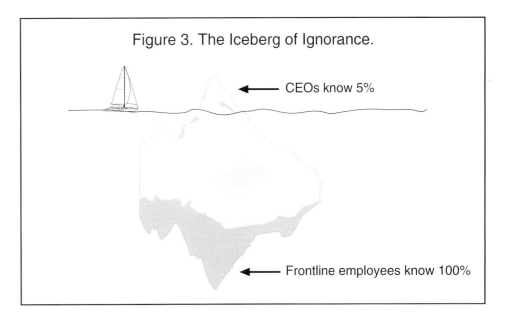

Figure 3. The Iceberg of Ignorance.

CEOs know 5%

Frontline employees know 100%

Quality Management expert named Sydney Shetafirm. Mr. Shetafirm says that CEOs know 5% of what goes on in an organization, represented by the tip of the iceberg above the waterline. At the other end of the iceberg are the frontline employees; they know 100% of what is happening but nobody asks them. They are the first to see the danger and the last to get the message. Conversely, if the frontline employees don't understand their connection with the big picture, they begin to have decreased response repertoire, which means they act like robots. They don't think anymore; they just do the job and they make mistakes.

To advance patient safety, we need to learn from high reliability organizations, who turn the Iceberg of Ignorance upside down. This means new skill sets and roles for the middle manager. Middle managers need technical skills and competencies, but they also need to assume the role of "broker" between the senior level and front line. For example, they need to translate the big picture (i.e., decreased reimbursements driving cost reduction) to the front line, and conversely, transport information from the front line back to the CEO in succinct terms so that the CEO knows what is going on in the organization.

Safe organizations are quality organizations

When Treasury Secretary Paul O'Neill assumed the role of CEO of Alcoa, he chose to focus on safety. When people questioned why he would focus on an area in which Alcoa was already very successful, he said, "If you have a safe organization, you have a cost-effective, quality organization because people have to understand their work processes." Safety acts as a proxy for overutilization and poor quality. If an organization organizes around quality, everything else will fall into line.

Moving the organizational focus from the sharp end to the blunt end

Healthcare organizations need to stop the focus on the "sharp end" as the culprit in healthcare error, and look instead at the system. At the sharp end is the poor individual who just happens to be in the wrong place at the wrong time. In safety science, the sharp end is defined as the vulnerable place in the system where errors appear; the place where the patient meets the provider. Despite their appearance at the sharp end, errors are due to multiple factors, not just one. For example, wrong-site surgery is usually blamed on the individual in the operating room, but maybe the X ray was turned upside down or lost. At the sharp end is the person who gets sued, blamed, and fired, but because the error doesn't get fixed, the very same error will show up again.

Active and latent error

An active error is the error that you see. It happens at the sharp end and is manifest in individual behavior—what a person does. In contrast, the latent error refers to the hole in the process that cannot be seen, but in reality produced the error. A good example is re-engineering nursing staff so that units have less-trained people and fewer nurses. There is nothing wrong with re-engineering except that it is done and people are often left to hang out to dry, without adequate and ongoing support. For example, there is no ongoing team training to teach a registered nurse how to work with a licensed practical nurse. Healthcare aides do not feel that they are part of the team in health care; often they do not feel able to talk to a nurse because they're not allowed to write on a chart. Errors can occur because of these holes in communication.

Work-arounds and diffusion of responsibility

Healthcare professionals are masters at work-arounds. Work-arounds are individually devised compensation patterns to accomplish a goal in a system of dysfunctional work processes. When work processes in the system are broken, individuals "work around" the system to figure out a way to complete the task at hand. Although work-arounds may make a job easier and even be the only way to get something done, unfortunately work-arounds also create an environment ripe for error.

Diffusion of responsibility means that people assume that someone else will do it: "If everybody is in charge, nobody is in charge." You may remember the widely publicized case in the 1950s in which thirty-seven people watched as a woman was raped and murdered. How could this happen? It happened because everybody thought somebody else was calling the police, a classic example of diffusion of responsibility. Applying the concept of diffusion of responsibility to an analysis of the errors in the hospital, I found that errors happened because when someone got busy, they gave up and assumed somebody else would finish the task.

Force function

Force functions are actions that make a system idiot-proof. A good example of a force function comes from your car; you can't start your car in reverse—you have to start it in park. This is a relatively recent innovation that was instituted for safety reasons. In health care, an example of a force function is taking potassium chloride off hospital units. It's more convenient for a nurse to have this solution on a unit,

Force function, continued

but 7,000 people die every year because potassium chloride is on the unit and is used by mistake. Most places have removed it, thereby preventing misuse errors.

Strength of changes for implementation

Changes to reduce error can be rank-ordered to show which changes will be the most effective. These changes, in the order of their strength (from strongest to weakest), are:

1. Force function,
2. Automation/computerization,
3. Protocols and preprinted orders,
4. Checklists,
5. Rules and double-checking,
6. Education,
7. Information.

If you look at how the list is prioritized, you'll notice that the "weakest" changes, the changes at the bottom of the list, are those changes that rely on actions by human beings. For example, checklists may or may not be filled out. Rules and double-checking are often forgotten if practitioners are too busy. Our favorite thing to do is provide people with more information, even though they are already on information overload. The more involved human beings are for that particular change, the less strength that change will have in affecting the process.

Level 3: Education

New skills are needed in the twenty-first century

The Institute of Medicine report released in March of 2001 includes a conceptual road map for health care in the twenty-first century, outlining sixteen chronic disease populations for health care to organize around. To serve these populations safely, new skills will be needed and training of healthcare professionals must undergo drastic change.

Revamping healthcare curricula

To begin to work on solving problems associated with patient safety, we need to provide tools to all of the people who are in training in this field—physicians, nurses, pharmacists, all practicing clinicians. The four components of safety science need to be integrated into the curriculum for all incoming professionals: instruction in human factors (outlined above in concepts such as sharp and blunt end and active and latent error), cognitive psychology, engineering (these concepts, outlined above, include force functions and work-arounds, as well as strength of changes of implementation), and team training. Given space constraints, only cognitive psychology is outlined here.

Cognitive psychology

Terms from cognitive psychology begin to provide a new language, the language of safety science, to describe and understand the events associated with human

Cognitive psychology,
continued

error in health care. Precise definitions of terms are needed to avoid misunderstandings. Three terms—slips, mistakes and near-misses—are provided here as examples.

The term "slip" describes impulse control errors. This is like putting shaving cream in your hair in the morning, instead of mousse. We know what to do; we just do it wrong. The best way to prevent slips is to impose a force function (i.e., move the shaving cream into another room). Slips are responsible for approximately 90% of the errors in health care, and certain working conditions (habit, frequent interruptions) predispose a person to make a slip.

A mistake is an error in judgment, such as putting a nurse on a chemotherapy unit without training in how to deal with those very dangerous drugs. Mistakes occur because we don't know what to do; we don't have the appropriate information.

Near-misses are jewels, what we want to collect and learn from in voluntary reporting systems. To know that change is being made in health care, we want to see a large increase in the number of reported errors, coupled with a simultaneous decrease in serious injuries to patients. For the culture of safety to become a reality in health care, the near-miss must become the treasure in health care, as it is in the aviation industry. An example of the aviation industry's lesson in the importance of tracking near-misses occurred when it was found that a plane crash at Dulles Airport would not have occurred had a near-miss been analyzed that had occurred three weeks before.

Level 4: Payers

Employers, insurers,
and managed care
organizations need to get
involved

Payers include employers, insurance companies, and managed care organizations. The Leapfrog Group, a Fortune 500 company, has taken the lead in harnessing the employer community to jump-start patient safety. The LeapFrog Group provides incentives in the form of preferential referrals to organizations who meet criteria based on three specific research findings known to enhance patient safety: (1) the implementation of computer physician order entry, (2) "intensivists" (specially trained and certified physicians) staffing intensive care units, and (3) sufficiently high volume for designated procedures. To date, insurance companies and managed care organizations have been relatively inactive in patient safety. One question that needs to be addressed is whether inappropriate denial of care is a safety issue.

Level 5: Alliances

Trade organization and
professional association
involvement

Professional and trade organizations also have a contribution to make to patient safety. The American Society of Health System Pharmacists and the American Hospital Association have made medication safety top priorities. The American Society of Health Risk Managers is focusing on disclosing error to patients. Last but not least, the American Medical Association had the foresight to create the National Patient Safety Foundation and continues to generously support its work.

Level 6: Governance

The environment must change

The government, regulators and accreditors, and the legal system all have important roles to play in patient safety, too. Legislation is pending to create a Center for Patient Safety within the Agency for Health Research and Quality, and to lay out the parameters for a voluntary and mandatory reporting system.

A word is in order about mandatory reporting of errors. In the truest sense, all reporting is voluntary. Mandatory reporting systems will not work if people feel that their livelihood is jeopardized by inadequate confidentiality protections. People will not report errors unless the environment protects them.

As of July 1, 2001, the Joint Commission has issued new patient safety standards, including a standard that mandates that healthcare professionals disclose errors. This standard is important, but professionals need to have training in order to have skills to disclose error. This becomes another competency to be addressed by education; no healthcare professional is trained how to break bad news in a way that will reduce the likelihood of litigation.

Level 7: Product Manufacturers

The interaction of products and people

Issues related to manufacturers center on pharmaceutical companies, makers of healthcare devices such as infusion pumps, and technology manufacturers. For a while, manufacturers did not see patient safety as their problem. Thinking that they make good machines or drugs that are safe and approved by the FDA, they had little awareness that the problems occur after products get into the marketplace and interact with human beings. A big issue in pharmaceuticals is look-alike packaging for drugs and sound-alike drug names. An example for device manufacturers is related to intravenous feeding and drug tubes fitting into similar couplers. A surgeon provided a picture that spoke to the role of product manufacturers in patient safety. The picture showed a catheter with multiple couplers inserted into the arm of a patient undergoing surgery. The surgeon said, "There will be an error reported, and I'm going to make it." Every one of the tubes on the catheter fit into the same coupler, making it very easy to connect the wrong coupler to the catheter. To date, there are no standards for technology, nor is it regulated.

Level 8: Societal and Cultural Influences

The role of the media

The media is one of several cultural influences that plays a very large role in the public's view of safety science. The media can be a help or hindrance. A well-known magazine published an issue on workers and health care, with a headline on the cover that read, "Too many jobs, not enough workers. How will we address the coming staffing crisis if we can't even solve today's healthcare labor woes?" The cover also included a fictional want ad that read, "Desperately seeking staff. Healthcare industry facing uncertain future needs hundred of thousands of frontline workers. Long hours. Difficult customers. Compensation competitive with the fast food

The role of the media, continued

industry. No stock options. If still interested, call 1-800-Tough-Job." This is an extremely negative image of health care and healthcare workers portrayed by the media. Would you want to apply for this job? Would you want someone who applied for this job to take care of you? Of course not. People who work in health care enter the field for good reasons. We have to reconnect to those good reasons about why they are there, and portray health care as an important, viable entity.

Another example of the role of the media is seen in recent articles on wrong-site surgeries and nursing that were very intelligently written, but the patient safety message was compromised by headlines that were sensationalistic. Magazine covers and headlines such as these do not help.

The Onion Model is not enough

The Onion Model demonstrated that managing processes, while critical in creating a culture of safety, is not enough to ensure patient safety. The entire system, including back-end factors such as the legal system, needs to work together to allow learning and improvement to take place.

A case example: Changing the system at Hermann Hospital

A root cause analysis at Hermann Hospital (Figure 4) showed that the biggest

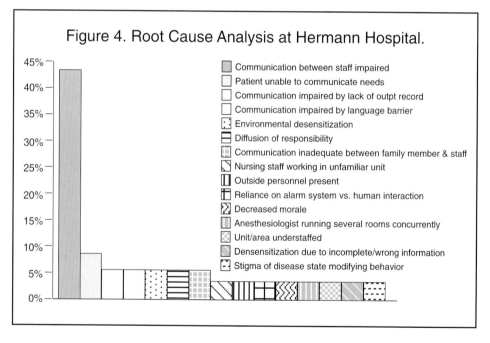

Figure 4. Root Cause Analysis at Hermann Hospital.

- Communication between staff impaired
- Patient unable to communicate needs
- Communication impaired by lack of outpt record
- Communication impaired by language barrier
- Environmental desensitization
- Diffusion of responsibility
- Communication inadequate between family member & staff
- Nursing staff working in unfamiliar unit
- Outside personnel present
- Reliance on alarm system vs. human interaction
- Decreased morale
- Anesthesiologist running several rooms concurrently
- Unit/area understaffed
- Densensitization due to incomplete/wrong information
- Stigma of disease state modifying behavior

predictor of error in our system was impaired communication between staff members; they basically didn't know how to talk to each other. So a performance improvement program was designed around communication.

We chose the area of medication-use process as the area that would most benefit from improved communication because of our sentinel event (the infant's death) and because the medication-use process constitutes the most common activity in an acute care setting. The medication-use process was divided into the four steps used in the JCAHO survey: ordering, dispensing, administration, and monitoring.

A case example: Changing the system at Hermann Hospital, continued

The first step was to determine which professional group was accountable for each step; that is, who owned each area.

In this model, physicians own the ordering process. The physicians' initial response succinctly stated the issue and offered a suggestion: "This is such a culture change to have to make this adjustment. Let's get big fluorescent stickers to stick on every medical chart," so we did. We requested that specific information be included on every sticker. For example, physicians were requested to print (rather than write) their name and beeper number on each chart. Physicians had to write the full name of the drug to be dispensed and spell out the microgram units. On pediatric orders, the drug dosage had to be calculated on the chart. House staff (people in training) needed to have someone co-sign high-risk orders (i.e., intravenous digoxigenin or vasoactive drugs).

All these changes sound simple to implement but ended up being very hard. For instance, nurses felt that they had to baby-sit physicians to write their name and beeper number on the chart, which made them angry and unearthed conflicts between doctors and nurses. (Nursing units even installed signs that said, "Got a name? Print it!…Got a beeper? Use it!") All of these things are documented in the literature as causes of error in the ordering process.

We did introduce a big culture change on our orders: a pharmacist was allowed to co-sign for a physician, reversing the hierarchy. This reversal came as the medical director reframed the role of the pharmacist for the physicians: "Pharmacists are a resource to your practice. They are saving you from getting sued. When a pharmacist calls you in the middle of the night with a question, you will thank them. You will not yell at them anymore for waking you up; you will thank them." It was a culture change and a statement of the medical director's courage.

One of the biggest days of fear in my life was when a Joint Commission surveyor said to me, "Tomorrow, I want you to show me how many medication errors you prevented last year." As it turned out, the pharmacists had a very sophisticated program but all of the data was sitting in a corner because they got yelled at when they called physicians to talk about errors.

Motivation to comply

Initially, Hermann Hospital had trouble getting physicians to comply with changes. So to create incentives, the hospital, which annually paid $10 million to the medical school for residency services, said that the money would be withheld unless physicians complied with the new medication policy. The medical school dean and department chair started encouraging physicians, and compliance increased.

The biggest success of this story is that at the end of the year, physicians' rate of compliance increased from 68% to 92%, but they themselves were not satisfied. They wanted 100% compliance in the next year. That was the true change; the people directly involved took the responsibility to improve.

Pharmacy changes

When we started monitoring the number of errors that occurred in the ordering process, we found that 33% were either sub-therapeutic or toxic, so we made changes to our pharmacy processes. (1) Consultations with physicians were strengthened, as stated above. (2) We limited drugs to a single strength. (3) We changed vendors to avoid look-alike drugs. (4) We color-coded look-alike drugs when vendor changes were not possible. And (5) we decreased the amount of information on labels to avoid confusion. All of these contributed to the reduction in medication errors shown in Figure 1.

Errors will be reported in the right environment

The culture can change. At Hermann, the first time adverse events were reported to the hospital's board of directors, our goal was to make errors transparent. Nonetheless, management was nervous. We were pleasantly surprised when board members, some of whom were astronauts, understood how frequently errors occur. As one board member said, "I'm really surprised you don't have more because this system is so complicated."

Upon hearing this presentation, a surgeon ran across the room and said, "You missed my errors. They weren't reported in your presentation." I realized right then that people *will* report errors if the environment is right.

The National Patient Safety Foundation

At the National Patient Safety Foundation, we work hard to serve as a catalyst to create a culture of safety in health care. The foundation is now four years old and was founded in partnership with the American Medical Association, 3M Corporation (an engineering company), CNAHealthPro (a payer), and Schering-Plough (a pharmaceutical company). NPSF is an independent, not-for-profit, multidisciplinary foundation with a single focus and a fifty-member board. NPSF is a proven convenor, which means we have been able to get diverse stakeholders together to deal with patient safety. For the first three years of the foundation's existence, NPSF's mission was to get patient safety on the national agenda. The 1999 publication of the Institute of Medicine report marked the end of this first phase.

The second phase, now that patient safety is on the agenda, is to serve as the connector between the world of research and regulation and the field of practice. NPSF maintains the world's largest, most comprehensive bibliography of patient safety literature. We publish a free *Focus on Patient Safety* newsletter quarterly, and an electronic newsletter twice a month called *Current Awareness* that encompasses the landscape of what is going on in patient safety. We maintain a web site and monitor a list server for the patient safety community, which will allow a physician in Great Britain to talk to a patient in the United States about issues.

For 2001 we are developing consumer communication tools and promoting a statement of principle on disclosure, and we published *Lessons on Patient Safety*, which uses the literature in the NPSF Clearinghouse and turns it into real-world lessons.

NPSF's Applications and Learning Program

One of our most active programs is our Applications and Learning Program. Our Solutions Awards Program gives $10,000 cash awards to frontline practitioners to recognize and disseminate patient safety solutions. Our National Patient Safety Consensus Projects convene groups for two days to come up with an agenda for patient safety in a complex area. The agenda-setting includes the development of a set of prioritized action steps with deep drivers.

We recently sponsored a Consensus Project on pharmaceutical safe use. Defining pharmaceutical safe use includes maximizing the benefit of a drug, minimizing the risk, and eliminating harm. The agenda we developed included (1) reframing drug safety as a national health priority, not as a regulatory issue; (2) developing a collaborative leadership model; (3) engaging consumers; (4) building public awareness about the risk of taking drugs; (5) educating healthcare providers; and (6) stepping up the qualitative process and policy research. An example of an action step coming out of this agenda is the "Think It Through Campaign," to engage consumers as active participants in safe pharmaceutical use.

NPSF's Research Program

In 2000, we received ninety letters of intent for our research grant programs, a 100% increase over 1999. These proposals cover areas such as pharmaceutical errors, educational interventions, organizational design, and error and incident reporting. NPSF's Research Program differs from other research programs in that we actively mentor our researchers. We also advocate for our investigators: if another agency will fund them, we will work with that agency to get them funded so we can fund someone else. And we get results fast.

We've also published a national research agenda and a catalog of research in the United States on patient safety. Both are available on our web site (www.npsf.org).

NPSF's Education Program

NPSF hosts regional forums that convene stakeholders in different states and regions of the country to talk about the issue of patient safety in a safe environment. We are building on these efforts to develop strategic partnerships at the local level. The goal of this effort is to have patient safety become a sustainable reality in different states. One of the big efforts of our Education Program is the development of patient safety curricula. In partnership with the American Society of Therapeutic Radiation Oncologists, we will create a model curriculum that will be used for other medical specialties and we are going to do an on-line multidisciplinary fellowship modeled after executive M.B.A. programs. We are partnering with Harvard University to do an executive session in Minnesota to educate CEOs, and we sponsor the NPSF Annenberg Conferences, known in the patient safety movement as pushing the patient safety envelope forward. For 2001, this conference focused on high-risk communication, and for 2002, the conference theme will be on creating a culture of safety.

Conclusion: Beyond process management to the systems approach

We need to move beyond process management. It's a critical tool, but we also have to have a systems orientation, rather than a linear one, to health care. This

Conclusion: Beyond process management to the systems approach, continued

means understanding and harnessing complexity. It means aligning all the agents that are acting in the system, which is no small task. It means reconnecting to our core business, or as one physician said to me, "retaking our vows." We need to have responsible and reliable systems of care. The answers do not lie in regulation or in technology. Rather, the answers are found in the systems approach, and are manifest in systems language: alignment and coalitions, collaborations and partnerships, integration and synthesis.

Health care can change, but everyone must work together to make it happen.

Author information

Joanne E. Turnbull, Ph.D., is a psychologist and social worker by training. Dr. Turnbull and her team at Hermann Hospital designed an error reduction program that not only achieved a significant reduction in serious medication errors but also transformed the organizational culture from "cover-up" to "accountability." The program drew national attention to methods of proactive medical error reduction and helped sound the call for transforming medicine's organizational culture to one of increased accountability.

Prior to her NPSF appointment, Dr. Turnbull was a senior healthcare administrator at Hermann Hospital, Memorial Hermann Healthcare System, and the University of Pittsburgh Medical Center. Dr. Turnbull has more than twenty years of experience as a clinician, researcher, and educator, developing expertise in change management and applied research. Her academic work includes over thirty research and clinical publications. She brings a unique perspective to patient safety: a focus on the complex systems and behavioral aspects of error reduction, bolstered by hands-on experience delivering health and mental health services and organizational improvement.

Dr. Turnbull received her bachelor's degree from the Pennsylvania State University, and received a doctorate and two master's degrees from the University of Michigan.

Editorial assistance for this article was provided by Daniel Picard.

Untangling the Web: Bringing Information Therapy to the New Healthcare Consumer

Author

Molly Mettler, Senior Vice President, Healthwise, Incorporated, Boise, Idaho

Introduction

Healthwise, Incorporated, is a not-for-profit consumer health information organization founded in 1975, with help from a Kellogg Foundation grant. Healthwise's mission is to help people make better health decisions. Healthwise believes that to have a better healthcare system, the role of the patient must first be re-invented. Our vision is to build a better patient by creating communities of the best informed, most empowered healthcare consumers in the world.

We've made great strides in twenty-six years. The organization struggled in the beginning because the consumer's role in health care had not historically received much attention. Self-care for patients had yet to become a prominent healthcare issue. With the advent of managed care, however, the reimbursement model changed and healthcare organizations became very interested in what consumers could do for themselves. As it turns out, consumers can do quite a lot.

Behavior of patients and demands on physicians evolve as information is more available

The way people behave as patients has changed. The twenty-first century patient is a "new consumer," with heightened expectations and demands. And, there is a tension between what patients want and what they are getting from health care. When patients visit a doctor to have medication prescribed or to consult about possible surgery, they come as consumers and they have certain wants that need to be met. Patients are increasingly looking for multiple things: information, relief from symptoms, positive outcomes, and empathy. The new consumer insists on choice, control, and customer service. If they don't get it, they're perfectly willing to shop around for a doctor, group practice, or health plan that meets their needs. Because of this new attitude, providers must reevaluate healthcare consumers' needs.

The roles doctors play and doctors' needs must also be considered if the healthcare process is to work smoothly. Doctors are challenged in this new era of the empowered and internet-active healthcare consumer and the emergence of a much more diverse and complicated healthcare system. Sure, doctors want to be paid, but beyond monetary concerns, they want to lead their patients to good health. They want to be able to use their skills, grow professionally, and feel like they are contributing.

When the requirements of both consumers and doctors are taken into consideration, then a system can be developed to effectively bridge what is now a significant gap between consumers' needs and the services health care provides.

Behavior of patients and demands on physicians evolve as information is more available, continued

The opportunity we see is to build systems that can bridge this gap. This gap between what consumers want and what they are getting can be bridged, we believe, through ensuring that patients and providers work in partnership, especially through the effective use of information.

Dealing with the challenge of a four-element crisis

Four simultaneous challenges are impacting the delivery of health care and the patient experience of that care. These challenges are:

1. The rise of the new consumer
2. The rise of evidence-based medicine
3. The rise of the internet
4. The call for quality improvement and patient safety.

These four challenges are shaping how health care will be delivered in the future.

The rise of the new consumer

The days of the passive patient following "doctor's orders" are over. Studies consistently show that North American healthcare consumers want information; they want choice in their health plan, in their physician, and in their treatment. They want to assume control. They don't want to feel like cookie-cutter patients; they want highly personalized health care.[1]

In a study published in the December 1999 issue of the *Journal of the American Medical Association,* only 9% of decisions were rated as informed decisions.[2] This means that millions of Americans, as consumers and patients, want something from a smaller group of healthcare professionals who are not equipped to provide it.

The rise of evidence-based medicine

It is astounding how differently medicine is practiced throughout North America; the mere factor of location can influence a patient's treatment. For example, a patient with lower back pain in Boston is likely to receive different treatment for this problem than a patient in Seattle. As illustrated by the *Dartmouth Atlas,*[3] the treatment prescribed for the same health problems varies all across North America. What the *Dartmouth Atlas* illustrates is that no one is certain of the right level of care. The challenge this presents is to ensure that health care is delivered upon the evidence and to make the scientific evidence available so that both physicians and patients can work together to come up with the right treatment plan for each individual patient.

The rise of the internet

The third challenge is the rise of public information through the growth and expansion of the internet.

A large population of new consumers is entering the readily available healthcare information marketplace; they want to be knowledgeable about their conditions and options for best treatment and positive outcomes. It used to be that a patient would only be able to talk with a doctor about the best way to address the patient's problem. With the rise of the internet this same patient can now go on-line and tap into a wide array of information about the problem.

The rise of the internet, continued

This presents a great opportunity for the consumer to make an informed decision. However, the patient will also have access to a lot of useless and potentially misleading information. Millions of internet users go looking for health information. The rise of technology has given the consumer the opportunity to gain more healthcare-related knowledge, but it does not yet offer effective ways for the consumer to evaluate all this available information. The challenge is knowing what is good quality information and finding a way to communicate it beneficially.

The call for quality and patient safety

A recent Institute of Medicine report found that an unprecedented number of people have died as a result of medical mistakes.[4] David Lawrence, president and CEO of Kaiser Permanente, one of the country's largest HMOs, after looking at his own system and others, said, "Our chassis is broken." The chassis of our healthcare system is indeed broken, and it will not be easy to fix. The chassis does not simply need a part replaced; it needs a new design.

Dr. Donald Berwick, president and CEO of the Institute for Healthcare Improvement[5] and a powerful advocate for quality health care, has indicated that the challenges facing the healthcare system are not marginal and the solutions are not incremental. Indeed, the challenge before us is a daunting one. And yet, we will never have a better healthcare system without first building a better patient.

Creating Information Therapy

Information is care. When good information is given to patients, patients become partners in their own care. "Good" information is characterized by five factors: (1) The information must be organized in a way that helps patients make good decisions; (2) It must be evidence-based; (3) It must be unbiased; (4) It must be referenced; and (5) It must be up-to-date. Good information rightly belongs in the hands of the patient. Good information can be prescribed, just like a medicine, a test, or a treatment, by means of "Information Therapy."

Information Therapy is *the delivery of specific information to a specific patient to better manage a specific health problem*. Think in terms of a pill being prescribed—when a patient has a problem that can be addressed by medication, the physician seeks to find the best medication for that particular patient and for that particular problem. In choosing the right medication, the doctor considers dosage, frequency, medium (pill or ointment), and many other factors.

Just like a pill, the delivery of Information Therapy also needs to be tailored to best fit the patient. The doctor must consider what is known about the patient and identify delivery methods that the patient will best respond to, be it one-on-one conversation, reading material, interaction with others, or personal research. The right dosage (or amount of information) must also be considered.

As with side effects to medication, the misuse of information must also be considered. The information a patient accesses must be carefully monitored. There are more than 25,000 web sites on the internet today that call themselves medical information or healthcare sites. Not all of these have been evaluated, but groups such as Hi-Ethics (Health Internet Ethics)[6] are working to certify medical information web pages.

The Healthwise Knowledgebase

Because the purpose of Healthwise has always been to get information to people to help them make better health decisions, we focused on what consumers said they wanted to know and we have embraced the qualifiers for good information listed above. To ensure that consumers have access to decision-focused, unbiased, referenced, up-to-date, and evidence-based information, Healthwise has been building an electronic consumer health information database called the Healthwise Knowledgebase. It contains information on:

- 1900 health topics,
- 600 medical tests and procedures,
- 500 self-help groups,
- 250 CancerNet topics,
- Medications (through an electronic drug reference), and
- 900 complementary medicine topics.

The Healthwise Knowledgebase is now accessible, under various names and in various configurations, to about twenty million families worldwide. The Knowledgebase represents our efforts to put quality information into the hands of the consumer/patient at the time of decision making. For a tour of the limited topics demonstration of the knowledgebase, please visit www.healthwise.org/kbase.

A study project: build a better patient

A model program, created and tested in Idaho and replicated across North America, is testing a simple idea: to build a better healthcare system, first build a better patient. Why "build a better patient?" Because educating and supporting consumers to take an active and informed role in their own care will yield the solutions that our current crisis clamors for. Historically, consumers' involvement in their own healthcare has often been overlooked by our more "formal" healthcare delivery system. However, let's look at where health care in this country is really practiced. Eight out of ten health problems that people experience are handled *by themselves, without the intervention of a healthcare professional.* What people do for themselves and their families, in their own homes, makes up the bulk of a hidden healthcare system, and yet this activity was largely ignored and unsupported until the advent of mass distribution of consumer self-care books, and nurse advice lines, to capitated populations by managed health care.

Research conducted over the past twenty-five years on medical self-care and shared decision making has shown, time and again, that when patients are informed and involved in their own care, three things happen: outcomes are improved, cost goes down and satisfaction goes up. In looking at quality improvement processes for health care, too much attention has been spent on limitations and restrictions on what doctors could or could not do—supply management. The attention needs to center on the consumers themselves—the ultimate demand creator for health care. If we truly want a better system we need to start where that system does—with the patient.

Healthwise set up a study project in Idaho with the goal of creating the smartest healthcare consumers in the world. With funds raised by a grant from the Robert Wood Johnson Foundation, local hospitals, and self-insured employer

A study project: build a better patient, continued

groups, Healthwise set out to create a community of informed and empowered consumers in a four-county area in southwestern Idaho. The study area covered about 10,000 square miles and included a diverse population, ranging from urban residents in Boise to rural residents in the surrounding mountain communities, with varying levels of education, familiarity with English, and access to technology.

In April 1996, *The Healthwise Handbook,* a 380-page guide covering 190 healthcare problems, was delivered to every residential mailbox in the four-county area. It gave information about symptoms, what to do to prevent or deal with problems at home, and what signs and symptoms indicate that a visit to the doctor is required.

First published in 1975 for sixteen mothers in a Boise community education group, there are now over eighteen million copies of this book in distribution worldwide. (For this project In Idaho, it cost $5.00 per book to customize, print, and deliver for 126,000 families. A Spanish language version was also available.)

Research has shown that when people get a book—not just *The Healthwise Handbook,* but any self-care book—and if they use it, they will save on doctor visits and emergency room visits.[7] In addition to delivering the handbook, Healthwise invested resources in educating the people in the community, teaching them self-care skills. They reached 13,000 people in the four-county area—holding sessions in church basements, grange halls, high schools, and anywhere people would gather.

In addition to educating the public, Healthwise also worked with doctors, nurse practitioners, physician's assistants, and medical office staff. These practitioners received training in different methods for increasing their patients' self-care and medical decision-making skills.

Resources were made available electronically to everyone in the study area through a restricted-access web site whose central feature was consumer access to the Healthwise Knowledgebase. For residents without personal computers, computer kiosks were set up in every public library and in many primary care physicians' offices as well as at large self-insured employers' work sites. In addition to web access and printers, these "Healthwise Information Stations" provided a variety of consumer health books in English and Spanish.

Healthwise went to great lengths to ensure that everyone had the opportunity to become an informed healthcare consumer. In addition to the web site and kiosks, there was the Healthwise Line, a nurse advice line. Nurses on this line gave general information, following the protocols of the Healthwise Knowledgebase, but did not give diagnoses.

The handbooks were sent out in April 1996. Six months later the Healthwise Line and the web site were activated. Almost immediately after the books went out, letters started coming in with stories about how the book had helped, and even saved lives.

It is important to emphasize that *The Healthwise Handbook* and the project are not structured to keep people away from medical care. Self-care teaches people when they need to go to the doctor. In some cases in Idaho, it got people to the hospital early enough to prevent further complications and save lives.

Outcomes

The Robert Wood Johnson Foundation awarded Oregon Health Sciences University (OHSU) a separate grant to evaluate what happened in Idaho against two control communities, one in Oregon and one in Montana. The university study found a reduction in visits to doctors in Idaho for conditions amenable to self-care, such as upper respiratory problems and minor muscular aches and pains.[8]

Blue Cross, looking at its own data, measured an 18% reduction in ER visits in the study area compared against two other Blue Cross regions in Idaho.[9]

Healthwise also reviewed how the handbook was received and utilized. Three years after the book was first distributed, OHSU researchers surveyed a representative sample of the households in the study area. Response to the handbook was positive. Nine out of ten households knew that the book had been mailed to them. Seven out of ten used the book at least once. In fact, the average use of the book was seven times a year. Four out of ten households reported saving at least one medical visit and two out of ten reported saving at least one ER visit.[10]

Physician response

From the consumer's point of view, through the Idaho study project, quality went up, desirable outcomes increased, and cost went down. Healthwise had succeeded in getting information to the consumer but was still uncertain how physicians would respond to their new empowered patients.

In times past, physicians were reluctant to make self-care and other types of information available to their patients because of concerns that patients would misuse the information and that there might be undesirable consequences. The results from physician reaction to the Healthwise Communities Project activities suggest that a significant change in physician attitudes has occurred over the years. OHSU investigated the use of the handbook and the kiosks in the physicians' offices. Ninety-one percent of physicians polled wanted the handbook in their offices for patient use. (However, only 75% of physicians reported referring their patients to the handbook regularly.)

Additionally, the Healthwise Information Station kiosks that were set up in doctors' offices were there only on a temporary six-month basis before being routed to another clinic. (Patients used the kiosks to find out information about the ailments they were seeing the doctor about, and also to research information about other health issues not related to the visit, such as sexual dysfunction and depression.) After the six-month trial period was over, all of the participating clinics said that they wanted to keep the kiosks in their offices. Time and first-hand experience with informed patients has changed many physicians' philosophy.

What's next?

By the end of the grant-funded period, Healthwise had met the study project's goals. The project *did* create whole communities in Idaho of smart, informed, and empowered patients, and physicians *did* endorse the program. Following this success at improving the quality of health care in Idaho, the project has been replicated in communities in Minnesota, South Carolina, Texas, and British Columbia.

What's next? continued

The Idaho-based Healthwise Communities Project was yet another step in the direction of building a better patient. The realization of true Information Therapy—being able to deliver targeted prescription information to a specific patient—is the next leap forward. As the following matrix shows, there are a number of important goals and opportunities for Information Therapy, and three basic tatics for accessing it.

Figure 1. Information Therapy Matrix.

Goals	Tactics	Opportunities
• Improve Quality	• Patient Initiated	• Prevention
• Reduce Costs	• Doctor Initiated	• Self-Care
• Increase Satisfaction	• System Initiated	• Self-Triage
		• Visit Preparation
		• Chronic Disease Self-Management
		• Shared Decision Making
		• End-of-Life Care

Innovative Information Therapy initiatives are looking at all the places where people touch the formal healthcare system, whether for self-care, prevention-oriented doctor visits, self-management of chronic illness, shared decision making for major surgery, or even end-of-life care. The goal is that every time a patient visits a healthcare practitioner and every time a patient addresses her or his own problem at home, the patient emerges from that encounter better informed and more empowered.

Information is essential to bridging the gap between what the consumer wants and what the consumer is getting from the healthcare system. Information Therapy allows doctors and patients to work together to find the best treatment. The mutual benefits of this are easily seen on the following chart:

Figure 2. Summary of Benefits from Information Therapy.

Patient Benefits	Physician Benefits	System Benefits
• Better Decisions • Involvement • Self-Management Skills • "Customer Service"	• Saved Time • Better Patient Compliance and Outcomes • Practice Enhancement • Increased Patient Satisfaction	• Improved Quality • Increased Capacity (Market Share) • Reduced Costs (Margins) • Increased Member Satisfaction • A Focus on Mission

If the patient is going to be the primary provider of care, it is imperative that the patient have as much information as possible. It is time to put decision making in the hands of the people who are most affected by it. Reinvent the patient, and by doing that, reinvent the healthcare system.

References

1. Kryouz, E. M., et al. *The Twenty-First Century Health Care Consumer.* Menlo Park, CA: The Institute for the Future, 1998.

2. Clarence H. Braddock III, Kelly A. Edwards, Nicole M. Hasenberg, Tracy L. Laidley, and Wendy Levinson. "Informed Decision Making in Outpatient Practice–Time to Get Back to Basics." *JAMA,* December 22/29, 1999–Vol. 282, No. 24, pages 2313–2320.

3. The *Dartmouth Atlas of Health Care in the United States.* Hanover, NH: Center for Evaluative Clinical Studies, Dartmouth Medical School.

4. Linda T. Kohn, Janet M. Corrigan, and Molla S. Donaldson. *To Err Is Human: Building a Safer Health System.* National Academy Press, 2000.

5. www.ihi.org.

6. www.hiethics.org.

7. Kemper, D. W., et al. "The Effectiveness of Medical Self-care Interventions: A Focus on Self-initiated Responses to Symptoms." *Patient Education Counsel,* 1992; 21:29–39.

8. Oregon Health Sciences University. *Idaho community health survey follow-up.* Portland, OR: Oregon Health Sciences University. Internal memo, 1998.

9. Sternberg L. *Healthwise handbook impact study.* Boise, ID: Blue Cross of Idaho. Internal memo, 1998.

10. Final Grant Report, Healthwise Evaluation Project. RWJF Grant ID#027929, May 1, 1996, to November 30, 1999. Merywn R. Greenlick, Chair and Professor, Department of Public Health and Preventive Medicine, Oregon Health Sciences Univ.

Author information

Molly Mettler, Senior Vice President, Healthwise, Incorporated, has been with Healthwise since 1985. A nonprofit organization, Healthwise is best known for the Healthwise Handbook, *now in its fourteenth edition, and the* Healthwise Knowledgebase *software. These products and the company's consumer education programs have won numerous national awards, including the American Health Book Award and the "Secretary's Award of Excellence," for a Distinguished Community Health Promotion Program, from the U.S. Department of Health and Human Services.*

In 1995, with funds from the Robert Wood Johnson Foundation, Mettler and her colleagues launched the Healthwise Communities Project. The vision: to make the 278,000 residents of four southwestern Idaho counties the most empowered, best informed medical consumers in the world. The project won the 1996 Spirit of Innovation Award and is being replicated in other communities around the world.

Mettler has authored scores of books and articles on medical self-care and health promotion, including Healthwise® for Life, *now in its third edition with nearly two million copies in distribution.*

Particularly passionate about health care for people age 50 and over, Mettler is Chairman for the National Council on the Aging (NCOA). She was the founding chair for the NCOA's Health Promotion Institute in 1990. That group honored her contributions by creating the "Molly Mettler Award" for leadership in health promotion.

Mettler has consulted for the World Health Organization, the American Association of Retired Persons, the Robert Wood Johnson Foundation, and Fortune 500 companies. She serves on the advisory boards of many organizations and is a Health Forum Fellow.

Editorial assistance for this article was provided by James Alphen, Victoria Fantozzi, and Laurence Smith.

Chaos Theory and Creativity: The Biological Basis of Innovation

Dr. Ary Goldberger, Beth Israel Deaconess Medical Center and Harvard Medical School, Boston, Massachusetts

Introduction

My organization is the Beth Israel Deaconess Medical Center where I'm the Director of the Margret and H.A. Rey Laboratory for Nonlinear Dynamics in Medicine—two names that may be familiar to you. Margret Rey and her husband Hans wrote the *Curious George* books for children. We try to emulate the spirit of Curious George in our laboratory—we get into places where we don't really belong. You may think that as a cardiologist I shouldn't be researching chaos and complexity theory or human creativity. Most people don't usually think that products of the imagination have much to do with biology. They are usually considered as separate disciplines: the arts and the sciences. They are in fact deeply melded. This article explores how chaos and fractal theory is relevant to health, to disease, to human creativity, and to what I call organizational pathologies.

Healthy individuals and organizations share the same three characteristics

As a physician I spend most of my time exploring the challenge of restoring and maintaining health. There are three defining characteristics of a healthy individual. 1. Productivity—the ability to do useful things. 2. Innovation—the ability to grow and change. 3. Resilience—the ability to bounce back from an injury or a setback; the ability to heal.

What are the defining characteristics of a healthy organization? When I took an unscientific poll at my own organization, I came up with the same list: a healthy organization is productive, innovative, and resilient.

Polyscopic analogy gives us insights to social systems

If we understand complex systems at one level, this may give us insights into systems at either lower (microscopic) levels or larger (macroscopic) levels. I call this the principle of polyscopic analogy. By coming to understand individual health, we may develop some insight into the health of human organizations and social systems. The connection is the new science of nonlinear dynamics, popularly called chaos theory. We can explore these relationships more thoroughly using these concepts and looking at actual data.

Natural systems are nonlinear

Chaos and fractals are interrelated parts of nonlinear dynamics. Almost all of classical mathematical physics and most every scientific field evolved from the assumption that the real world can be modeled using linear equations. But linear equations and models are, at most, a first and very rough approximation of natural systems. Nonlinear systems don't behave according to the classical rules. They do wild and weird things—they are the real world.

Fractals are the order within chaos

If you measure something that is chaotic in the technical, nonlinear sense of the word, it dances around in a way that looks completely irregular. However, if you re-graph it in what's called a phase plot, you find that a system that first appeared random as a time series may indeed have remarkable, hidden structure. It now appears organized in a complex way (Figure 1).

Nonlinear systems are doubly counter-intuitive. Irregular dynamics like those shown in Figure 1 can arise from very simple systems, provided they are nonlinear. You can write a set of rules and the system may still explode in your face. It can behave in a very ill-mannered, unpredictable way, just because it's nonlinear. Chaos is defined as the irregularity that arises in systems that appear relatively simple, but are nonlinear. Equally remarkable is the fact that hidden in this apparent disorder is a type of architecture, a hidden order that relates to fractals. Fractals are the geometry, or the order within chaos.

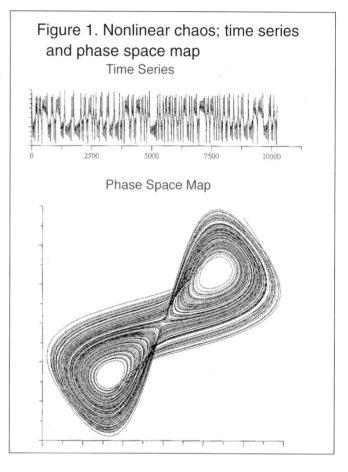

Figure 1. Nonlinear chaos; time series and phase space map

Time Series

0   2500   5000   7500   10000

Phase Space Map

Fractal patterns have scalar variance

A fractal is a pattern that repeats itself on various scales. It is composed of sub-units and sub-sub-units that resemble the larger structure. A tree is fractal: it has branches that have smaller branches that have twigs (Figure 2). Another example is the Russian matreshka dolls: hollow wooden dolls nested within dolls, a folk art image thought to symbolize fertility—life within a life. In the US, both federal and state court systems are organized in a fractal-like branching network. The lower courts feed cases to the higher courts, like streams feeding into larger rivers. This sort of property is called self-similarity, or scale-invariance. A fractal then has no charac-

Figure 2. Fractal branches

Fractals have scalar variance, continued

teristic size. For example, there is no single length of a tree branch. You might say a branch is this big, but that's only one branch. A twig or a limb is smaller or larger. So, branches don't have a single length: they have a wide range of lengths.

Importantly, this concept not only applies to geometric structures, but also to processes. Processes can be fractal when they incorporate different time scales; the fluctuations over different time scales resemble each other. What is the characteristic temporal scale of human life? Some biological processes take microseconds or seconds. There are other processes that take months or years. All of these processes occur simultaneously, in a symphonic way. Fractal processes do not have a characteristic temporal scale. I'll come back to that because that turns out to be basis of much of healthy physiology.

Fractal designs have many advantages

Fractal designs are very robust and redundant. These are very useful structures to diffuse nutrients throughout a system. Branching systems are very information-friendly: they allow information to percolate throughout the network. This becomes very relevant in physiology and social systems.

Fractal design infuses artistic creation

People like to look at fractals; many quilt patterns are fractal (Figure 3). Another example is the work of the great Japanese artist Katsushika Hokusai. In Figure 4, we see waves with Mount Fuji in the background. There are big waves that have smaller waves, that have smaller waves, that give off smaller wavelets. Mount Fuji in the background is also fractal. The closer one looks at a mountain the more irregularity and structure is revealed. The remarkable thing about this work is that if you were to move the image of Mount Fuji to one of the waves, and put the image of the wave where Mount Fuji was, you couldn't tell the difference. Hokusai somehow recognized the correspondence—the equivalence of the form of the waves and the form of Mount Fuji. Nature seems to favor certain forms in very different settings: we call this correspondence *universality.*

Tom Stoppard

Figure 3. Fractal quilt design

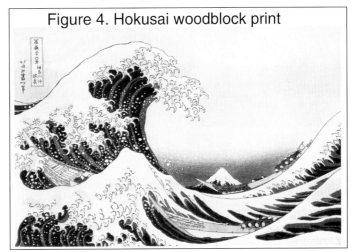
Figure 4. Hokusai woodblock print

Fractal design infuses artistic creation, continued

wrote a play called *Arcadia* that is remarkable because it's a metaphysical exploration of fractals and chaos theory—in fact, Stoppard gives a glossary of terms in the playbill. Stoppard's play explores explicitly what artists and writers have always used implicitly in their compositions: the harmony and rhythm of fractal design, and the excitement and drama found in the dynamic movement of chaotic, nonlinear systems.

Figure 5 shows a glorious example of fractal patterns in architecture. The Gothic design of the Cathedral of Cologne is made up of structures that have smaller structures that have yet smaller structures–arches upon arches, crenellations upon crenellations, spires upon spires. That's a very fractal-like form.

Figure 5. Cathedral of Cologne

Fractal design is inherently artistic

Perhaps when we walk into a cathedral, we are literally entering a re-creation of nature. It's a moment of self-discovery because these cathedrals are built using nature's fractal architecture, reminiscent of the branching patterns of trees in the great forests, the wrinkliness of mountains, and the fractal patterns found in our bodies. We are fractal machines, and by entering a cathedral we are, in a sense, entering ourselves. The great architects who designed these were unconsciously projecting human physiology and natural forms on a macroscopic scale.

Fractal lines are irregular and "wrinkly"

Fractal lines are wrinkly like the one shown in Figure 6. They look like coastlines. If you magnify a fractal line, what you see are wrinkles upon wrinkles, and if you magnify those wrinkles, you once again see wrinkles upon wrinkles upon wrinkles. If you look at a coastline from a distance, it looks smooth and regular. The

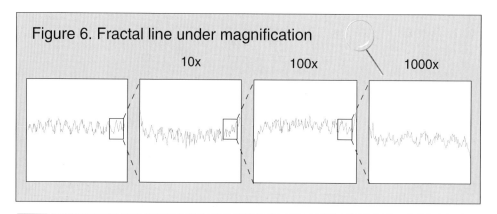

Figure 6. Fractal line under magnification

10x   100x   1000x

Fractal lines are irregular and "wrinkly"

closer one looks the more and more irregular it becomes.

Fractal geometry differs from the mathematics we grew up with, which may be why so many of us hated high school geometry—it seemed so irrelevant to the natural world. As Benoit Mandelbrot, the father of fractal geometry, has described, nature doesn't produce straight lines, rectangles, and cubes. Instead, the real world is filled with irregular and crinkly shapes, and branching structures. That's fractal geometry, and we need a different type of mathematics to describe it.

Fractional dimension

What is the dimension of a line like the one shown in Figure 6 that has wrinkles upon wrinkles? It is not one-dimensional because it has more structure. But it doesn't fill up a plane, so it is not two-dimensional. The dimension of a wrinkly line like this is somehow between one and two. It's an in-between or fractional dimension, which is where the term *fractal* comes from. It refers to the things that exist in between the dimensions of classical geometry.

Fractals, therefore, have non-integer dimensions. We measure lines using a ruler. But if you try to measure a fractal line, such as the one in Figure 6, the length will depend on the size of the ruler (Figure 7). If you plot the length of a linear line against the size of your ruler, the result will always be the same. You can use a big ruler, or a small ruler, you'll get the same length. But if your line is a fractal, the smaller your ruler, the more the detail you will pick up, and the longer your measurement becomes.

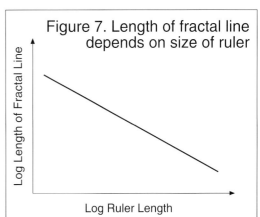

Figure 7. Length of fractal line depends on size of ruler

Log Length of Fractal Line

Log Ruler Length

Suppose you measured the coastline of Great Britain using different methods. The first measure was made using a picture from a satellite. Next, if you walked the coastline—you would measure a much longer distance, because you'd be walking along that entire craggy coast. But suppose you put a pedometer on an ant's foot, and had it trace out the coastline. By the time it got back, it would measure a very long line, because the ant is going to pick up even more of the wrinkly detail. Trees don't have a characteristic branch size, coastlines don't have a characteristic length, and fractal processes don't have a characteristic temporal scale.

Normal heartbeat is fractal

The geometry of fractals and the inherent strength and advantages of fractal designs leads us to physiology and to your heartbeat; and on a more macroscopic scale, to health and disease. Pictured in Figure 8 on the following page are records derived from an EKG of a healthy person. Although your pulse normally feels quite regular, if you very precisely measure the interval between heartbeats, a surprising finding appears. The normal heart rate is not metronomically regular, even during sleep. Instead, there is an imperceptible but highly complex variability in the normal

Normal heartbeat is fractal, continued

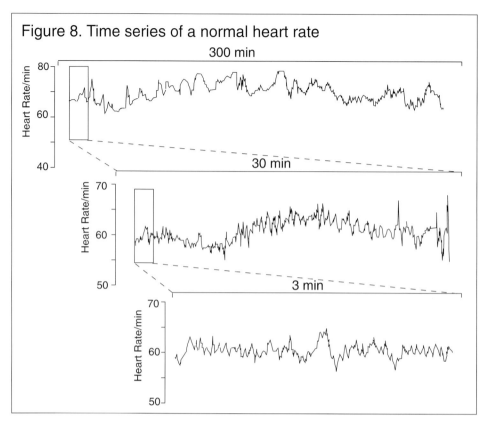

Figure 8. Time series of a normal heart rate

heartbeat from one instant to the next. It doesn't settle down to a constant rate, as predicted by the classical theory of homeostasis. Instead, it jumps around in a way that's quite unpredictable, even under the most placid circumstances.

The time trace of a normal heart rhythm has a fractal, "wrinkly-coastline" irregularity. We can prove this in a more rigorous way, but, put simply, the heartbeat is fractal because, measured over several hours, the heart rate doesn't settle down. It moves around, jumping up and down. If you look over a smaller scale, let's say thirty minutes, the heart rate still traces this irregular coastline-like portrait. Even over a three-minute span, it doesn't smooth out. That's the nature of a fractal process: its behavior is not regular and smooth, and it shows irregular variations over multiple time scales.

This type of irregular fractal "coastline" plot not only describes your heartbeat, but is also mathematically comparable to the fluctuations of pitch in classical music. By comparing the two mathematically, we found that the frequency spectrum is essentially identical to that of classical music.

A biological basis for musical invention

In the spirit of Curious George, we thought that we could do a little experiment. We started with the heartbeat intervals and converted them into a sequence of musical notes. But we needed someone to add an accompaniment to these. Fortunately, there was someone around my house who was able to do that at no cost—my son. Zach David Goldberger is a professional musician who performs under the name Zach Davids. He is also a pre-med student, so it was doubly convenient. To his surprise when he started to play these notes generated by heartbeat variations, he

A biological basis for musical invention, continued

found that the notes actually sounded musical. So without much trouble at all, he was able to compose music around these heartbeat-generated notes, and create what we call "Heartsongs: Musical Mappings of the Heartbeat."

"Heartsongs" are complex physiologic rhythms transposed onto a musical diatonic scale with some embellishments. If your heartbeat were perfectly regular, smooth, and metronomically pure, it would make for very boring music. If your heartbeat were completely random, it would sound like some kid pounding away at the piano. Was the result of our experiment plausibly musical? Any answer is going to be a subjective one. But most people who listen to the songs agree that they have musicality. Some people find them particularly relaxing. Classical music and healthy heart rate variability share a fractal connection. Both are products of complex nonlinear control mechanisms. [A sound clip of "Heartsongs" appears on our website, www.goalqpc.com.]

Let me pause and give proper credit—we are not the first to observe the connection between music and the heartbeat. More than four centuries ago, William Shakespeare wrote, "Ecstasy! My pulse, as yours, doth temperately keep time and makes as healthful music." *Hamlet* (Act 3, scene 4)

Artisitic creativity represents an externalization of biological dynamics

An article written about our work that appeared in *The New York Times* [1] a while ago described our speculation that creativity in some way represents an externalization of biological dynamics. Artists such as Hokusai, composers like Bach, and the architects of the great cathedrals projected onto a canvas, onto a musical score, and into the design of the buildings, very deep innate biologic rhythms. When we hear, see, or experience these works of art, these are in fact moments of self-discovery. We are seeing and hearing patterns that are already inside of us. And when it is done just right, there's a magical resonance. Our Heartsong "experiments" may help demonstrate this biological connection to artistic creation.

Disease: the breakdown of fractal patterns

What is the result if something terrible happens to a biological system that has this fractal "playfulness?" Perhaps the most common thing that happens is that the system loses its variability, and becomes pathologically regular: the time scale repeats itself again and again. When systems become excessively regular, there is an increase in predictability and a loss of resiliency. Figure 9 on the next page contrasts the excessive regularity of a pathological heartbeat with the complex variability of a healthy heartbeat.

Warning: excessive periodicity is bad for your health. Seizures can be induced in susceptible individuals by giving them excessive periodic stimulation with light, called stroboscopic stimulation. There was recently an epidemic in Japan after children watched a cartoon character repeatedly flash its sparkling eyes—over 500 children were admitted to emergency rooms, some with apparent seizures. The human brain doesn't like to be excessively and periodically stimulated.

The second way the system can break down is to become completely random. It traces a stochastic signal, such as white noise or static. An example of this occurs when the heartbeat becomes completely irregular during the arrhythmia called atrial

Excessive periodicity or random behavior, continued

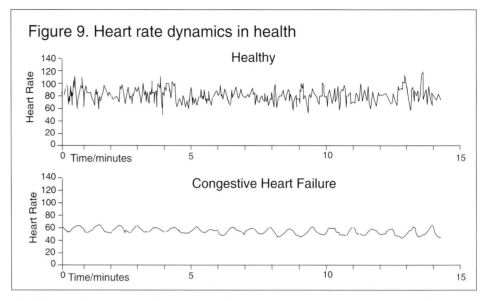

Figure 9. Heart rate dynamics in health

fibrillation. The uncontrolled growth of cancer cells may be another example.

The paradox of clinical diagnosis

Pathology correlates with a loss of fractal complexity. We have evolved the concept of disease as a *decomplexification* of a system. Counterintuitively, the output of many biological systems becomes more regular and more predictable with pathologic perturbations. Although we call pathology in systems *disorders*, many diseases manifest patterns that are highly ordered and periodic: the tremors of Parkinson's disease, manic depression, autism, obsessive-compulsive disorder, brain waves during epilepsy, and breathing patterns in heart failures. In fact, clinical medicine is not feasible without this paradox; without such stereotypic periodic behaviors, clinicians couldn't make diagnoses.[2]

The aging process is also typified by a loss of complexity. A comparison of the cross-section of the brains of elderly persons and young accident victims shows a much richer pattern of neural branching in the latter.

Healthy function is difficult to characterize

Healthy function, on the other hand, with its broadband, fractal-like variability, is much harder to characterize. It's very hard to use a single word to describe healthy behavior. We end up using words like plasticity, variability, resilience, and productivity. This type of variability enables the organism to adapt.

Human organizations have fractal structure

If, like Curious George, you have followed me this far, let me suggest that fractal theory may be applied in polyscopic analogy, not only to human biology, but also to human organizations. There are three defining features the "3F's," which underlie productivity, resilience, and innovation, the hallmarks of healthy functioning individuals and organizations.

• Fractality—branching, treelike complexity.

• Feedback—interdependence dominates interactions; the interactions are nonlinear.

• Ferment— systems don't sit still, and are most stable when driven far from equi-

Human organizations have fractal structure, continued

librium. These systems jump around; they don't settle down to a constant steady state.

The corporate analogs to corporal diseases which degrade these "3 F's" are—

Corporal Diseases	Corporate Diseases
• Excessively periodic behavior	• Excessive rigidity or over-regulation; too much predictability.
• Random behavior	• Anarchy, hierarchical dissolution.

Healthy systems, therefore, live in between complete randomness and excessive order—that healthy zone is fractal.

New parameters for testing an organization's health

This raises the possibility that we can do new types of diagnoses on organizations because we can potentially test for these features. I can't give you a specific test yet, but I would suggest the following parameters for measuring corporate health:

1. Draw an organizational chart of your company. Does it have a treelike appearance? Are there nested hierarchies? Fractals disseminate information very quickly. Does information percolate efficiently through your system? Or do you have a bunch of middle managers running around in circles, talking only to each other?

2. Is your organization auto-inventive? Does it readily evolve new structures to adapt to unpredictable challenges on different time scales? Is it playful? Is it creative? Is it innovative? How varied is the repertoire of your organization? How much spectral reserve does it have? Does it resemble a composition by Bach? Or is it a one-note marching band?

Managers are really choreographers, composers

How can we cure sick organizations? I suggest to you that rather than thinking of ourselves as *managers*, we should think of ourselves more as choreographers, composers, and conductors. And by doing so, we can restore the music of the heart not only to ourselves, but also to our organizations.

Note: This article was based upon a presentation Dr. Goldberger made at the 1998 GOAL/QPC Annual Conference.

References

1. *New York Times*, 31 October, 1995.

2. Goldberger A.L., "Non-linear Dynamics for Clinicians: Chaos Theory, Fractals, and Complexity at the Bedside." *Lancet* 1996; 347:1312-1314.

Author information

Dr. Ary Goldberger is a graduate of Harvard College and received his M.D. from Yale Medical School in 1974, where he was an intern and resident. He received his cardiology training at UCSD. He is currently Associate Professor of Medicine at Harvard Medical School where he is the Director of the Electrocardiography and Arrhythmia Monitoring Laboratories at Beth Israel Deaconess Medical Center, as well as the Director of the Margret and H.A. Rey Laboratory for Nonlinear Dynamics in Medicine. Dr. Goldberger's pioneering work in the application of nonlinear dynamics in medicine is widely cited.

The Future of Medicine

Author

Andrew Weil, MD, Director, Integrative Medicine Program, and Clinical Professor of Medicine, University of Arizona Medical Center, Tucson, Arizona

Editor's note

This article is based on an address Dr. Weil gave in the Celebrate Life *speaker series by the Center for Mindfulness in Medicine, Health Care, and Society, University of Massachusetts Medical School, Worcester, Massachusetts (www.umassmed.edu/cfm). Dr. Weil spoke on the deepening crisis in health care, the widening gap between the public's expectations of doctors and the realities of medical education, the emerging field of integrative medicine, and the development of new medical and scientific paradigms—*Editor.

Introduction

Medicine is in terrible trouble in this country. Conventional medicine has become too expensive and is threatening to sink our healthcare system. But at the same time, there is a movement that has the potential to redefine our understanding of medicine and healing. By enlarging and expanding the paradigms of conventional medicine, practitioners have the opportunity to become better healers and to better serve their patients while revitalizing the healthcare industry.

Nutrition

Let me begin by saying something about nutrition, food, and medical research. There is more confusion about nutrition in this culture than at any time in my life, as witnessed by the proliferation of completely crazy diets and recommended ways of eating that are irreconcilable with each other. You have people telling you to eat as little fat as possible. You have people telling you to eat as little carbohydrate as possible. There are even people telling you to eat according to your blood type, which makes no sense at all to me. Things should not be going in that direction; it should be getting less confusing rather than more confusing.

Interpreting "research" and understanding its limitations

This confusion is representative of a general problem that we have of just too much information. It is impossible for people to interpret all of the information that comes at them these days. What you're being forced to do, in a sense, is to rely on authorities or experts who can interpret information. In addition, there is very little training in this culture, both within medicine and outside of medicine, on how to interpret research findings.

Not all information that comes out of scientific research is worth paying attention to. There are a lot of badly designed studies. Many studies answer questions that they were not set up to ask. The conclusions are invalid. The research methods might not have been right. And yet these findings get plastered all over the newspapers, adding to our confusion.

I'll just give you a couple of examples. Early in 2000 there was a widely publicized study that maintained taking supplemental vitamin C thickens coronary arteries. The findings of this study have simply become accepted as fact. I remember being on a television show in California. The cameraman saw someone

Interpreting "research" and understanding its limitations, continued

taking vitamin C and responded by saying, "Oh, you can't take vitamin C anymore. It's bad for your arteries." Suddenly, the idea that vitamin C is detrimental has become accepted as fact. This was a random finding from an unpublished study that is inconsistent with everything else we know about vitamin C; it contradicts all of the information about vitamin C's antioxidant properties, which protect against cardiovascular disease. This study simply highlights a finding, unsupported by any other evidence; it has no clinical significance at the moment.

Problems associated with retrospective studies

There was another study done about tofu causing dementia. It was actually an interesting study that was published in a good journal of nutrition that studied the eating habits of a large number of Japanese-American men in Hawaii. It concluded that those Japanese-American men who ate the most tofu in middle age now have the highest rates of dementia. So I've been getting all sorts of letters and inquiries from people asking questions: "Is it safe to eat tofu anymore?" "Does tofu cause Alzheimer's disease?" Well, this is an interesting finding, but like much research in this area of food and health, it's what's called a retrospective study. A retrospective study involves looking at an effect in the present (dementia) and trying to tie it to a cause in the past (eating tofu in mid-life). The only thing that retrospective studies can do is raise questions. They may suggest hypotheses that then have to be tested. This is a logical fallacy known as "After the fact, therefore because of the fact." The problem is that an effect you see in the present could have any number of causes in the past. But to try to single out one thing and say that it's the cause is very risky logically.

One possible alternative interpretation is that Japanese men of this generation who ate a lot of tofu in mid-life and are now in their late seventies and eighties were probably the Japanese who adhered most closely to traditional lifestyle. Therefore, these men might have had less access to good medical care. That's just one alternate explanation. There could be any number of reasons that might explain why their health in late life would be different from those of other Japanese.

At any rate, all you can do in a retrospective study is raise questions. Then you have to set up prospective research. For example, you might divide a group of population into two groups and feed one of them a lot of tofu, watching them over time to see if there are any differential rates of developing any problems. Prospective studies are harder to do; they're more expensive, and they're more time-consuming. Most of what we see in the area of relating food to health is of the retrospective sort, which is very shaky logically.

Trends in nutrition and eating

I see trends in eating in this culture going in two directions at once. On one hand, we have much better food available to us than we did when I was growing up. Certainly, there is much better food available in many grocery stores in most parts of the country. There is much more availability of organic produce. Food in restaurants has evolved. We have much greater choices due to the variety of ethnic cuisines that

Trends in nutrition and eating, continued

have come into our culture. All that's terrific.

At the same time, I think there are more and more people eating more and more fast food and processed food. The eating habits of kids have gotten worse and worse. Both of these trends are developing at the same time.

Difficulties in making changes associated with nutrition in institutional food service

In the Integrative Medicine Program at the University of Arizona, we've done many things that I think are quite radical and are really a stretch for a university medical center. We've been bringing in energy healers to teach physicians how to sense energy and transmit energy. We've been teaching physicians about quantum theory and chaos theory. I would say that's pretty radical stuff, and we've been able to do it. What we haven't been able to do, or make the slightest progress at, is changing the food in the hospital cafeteria. This is really part of the trend toward eating more fast food and processed food, which is difficult to reverse. In many cases, it has become institutionalized.

I had no idea how difficult it would be to improve the food in the cafeteria. In 1999 there was a big announcement that the university cafeteria was going to be remodeled and that it would be shut down for two months. I thought this would be a great opportunity to propose some things that could be done differently. It took a great deal of effort to get a meeting set up with the powers-that-be that ran the university cafeteria.

As a side note, I thought that the stumbling block was the registered dieticians who were in charge of food selection. In the past I have blamed registered dieticians for a lot of the food problems that we have at institutions. But in fairness, I have since learned that even if dieticians wanted to do better, they couldn't because they are at the mercy of the food-services people with whom the institutions contract. These are huge companies that exert a dominant force on almost all the food served in institutions in this country. So it was with the agents of one of these firms that we met, and they were not a cordial group of people.

Their only interest was profit. Any change we suggested was opposed, based on their fear that it would lower their profit. For example, the one section in the old cafeteria from which I could get something to eat was the salad bar, and it was a pretty minimal salad bar. We started talking about the possibility of expanding the salad bar. They immediately said that their preference would be to eliminate the salad bar because it was a money loser.

Canyon Ranch, the big spa in Tucson with whom I work and our program works, offered, gratis, to furnish a fabulous salad bar for the university cafeteria. Well, the food service contractor would have none of it. After much argument we looked at the catalogs and asked them to put in a bigger salad bar than they had planned on. Eventually they agreed to it, or at least we thought they did. But at each subsequent step, the salad bar was downsized. It became the case of the incredibly shrinking salad bar, and we wound up with one smaller than the original one.

We also proposed serving one dish a day that was recommended by the Integrative

Difficulties in making changes associated with nutrition in institutional food service, continued

Medicine Program. We would give them recipes and just have one dish a day where they would post a sign saying it was recommended. No, they wouldn't do that. We asked them if they would provide nutritional information for the food so customers could see the fat content. No, they wouldn't do that either. So here we are with the same food, or food that is worse than ever, despite all of our efforts to improve it.

Lack of nutritional choices

There are other areas like this in which I feel very victimized in our country. I definitely feel that way in airports. When I was travelling recently, I had to change planes, as I often do, where you can walk for miles past endless food outlets and it's all the same stuff. Why do we put up with that?

Why do we put up with the kind of food that's being served to kids in school cafeterias? The trend in our country now is to let fast food chains become the food providers in school cafeterias, and people think this is a great idea.

Another area that's horrendous is the food that's served in the institutions for the elderly in this country. This is a time in life when nutrition is especially important, when many people want to eat lighter food and more nourishing food, and the food served, again by these big food-service providers, I would characterize as 1950s retro-cuisine. It's pork chops with cream gravy and the heaviest desserts imaginable. Why do we put up with this? What will it take to have things change in this area?

Show people that healthy food can be appealing

We need to show people that healthy food can be delicious. Because all of us have been served really dreadful food while being told that it's healthy and good for us, a lot of people have created in their mind a conflict between food that's healthy and food that's enjoyable. It doesn't have to be that way.

In addition, a lot of people think that preparing food is a huge and difficult chore. People have to be shown that food preparation can be easy and fun. A very good strategy is involving kids in food preparation from a young age. I think that's a way of creating interest in making food and building better eating habits. Alice Waters, the famous chef in California who started Chez Panisse restaurant, is a major champion of organic food. She started a program of getting organic food served at schools in the Bay Area in California, which is great. She's done work in disadvantaged areas of Oakland and Berkeley where they've started community gardens and children growing their own food. So there are ways of doing this.

But we should not kid ourselves about the forces we're up against. There are very powerful economic interests that want to see Americans eat junk. And whether that's in homes for the elderly, or school cafeterias, or airports, those are powerful interests and they are really catering to our inherited tastes for sugar, meat, and fat, and our learned taste for salt. A lot of education needs to happen, as well as practical demonstrations, to show that you can make food that's quick and easy to prepare, delicious, and also healthy.

Lack of nutritional knowledge

I don't think the quality of food is going to change unless consumers get angry enough and aroused enough to demand something different. One reason why things are the way they are is that our doctors are so inept in the area of nutrition. They are simply non-forces in this area. An informed medical profession could be a very powerful catalyst for change, but the problem comes back to how doctors are trained. Nutrition is not part of the medical curriculum. Occasionally when I say this, medical school faculty or deans object, but I will stick by what I say. The total instruction in nutrition that I got in four years at Harvard Medical School and one year of internship was 30 minutes, which was grudgingly allowed to a dietician at the Peter Brent Brigham Hospital to tell us about special diets we could order for patients. That has not changed substantially since I've been out of medical school.

There are 20 percent of schools that now say they teach nutrition, but when I look at what they teach, it is not nutrition in my book. It's mostly bio-chemistry. It does not prepare doctors to answer questions like, "Is it safe to take beta-carotene as a supplement?" or "Is it better to eat margarine or butter?" Doctors don't know the answers to those questions unless they have made an effort to learn this information on their own.

The fact that doctors are so poorly educated in this area is one reason why the food is so awful in hospitals and hospital cafeterias. It's also why we can't make any headway against the for-profit forces that have dominated these institutions and determined their food choices. This lack of education on nutrition is one glaring deficiency in the way that doctors are trained.

But there are also many other glaring deficiencies. The more that I think about it, the more I am convinced that until we change medical education and the way doctors are trained, we will continue to head in the direction we are going. There is a widening gulf between what consumers want from medical doctors and the realities of what medical schools are training doctors to do.

State of health care

The things that I have mentioned so far merely hint at the problems in medicine today. It's no secret that medicine is in terrible trouble in this country. In fact, our entire healthcare system is. There is an accelerating rate of bankruptcy among hospitals and clinics. This is a trend that I feel will continue. I can easily envision a time in the not-distant future when large areas of our country are going to be left with only one central hospital, which will be the only one that can afford all the hardware.

For many doctors, this economic crisis seems to have come out of the blue. But I think it was very predictable. As far back as 40 or 50 years ago, one could see these trends developing. The bottom line is that conventional medicine has become too expensive. It's become too expensive for a number of reasons. Some of them are within our control, and some of them are not within our control.

A couple of reasons that are not within our control have to do with our success. In the early part of the century, we rolled back infectious disease. As a result, we're left with chronic degenerative disease, which is a much harder and more

State of health care,
continued

expensive thing to treat. In addition, we have enabled people to live longer. Our population demographics are changing significantly. Population experts say that we should start bracing ourselves for what they call the coming demographic bulge of elderly people as the baby boomer generation gets into its senior years. As you know, the greatest medical costs are in the last years of life, so we haven't seen anything yet. We're just at the barest fringes of this change in population demographics that is going to impact medical costs.

One reason that healthcare costs are so high that is within our control has to do with the nature of medicine and people's expectations of it. Conventional medicine has become increasingly dependent on technology. The technology is inherently expensive. There's no one way around that. This enthusiasm for technology probably began a little before the beginning of the twentieth century. It produced great results in medicine throughout the first half of the twentieth century and great advances in our ability to intervene in instances of disease.

Technology and its complications

In medicine, as in other areas of society, we are seeing that technology creates as many problems as it solves. The particular problem that's been created is expense. It's the major reason the bills can't be paid anymore. This is not something we can think our way out of. It's not a matter of tinkering with the mechanics of reimbursement to make it work again. The system is failing, not just in the United States but all over the world. Everywhere that technological medicine has grown dominant, it is now beginning to decline, causing the insurance systems in these countries to fail, among other problems. The situation in the United States could be worse.

Japanese medicine

Let me give you a glimpse of medicine in Japan, which is a country I know well because I go there frequently. The Japanese have taken healthcare ideas from our culture and used them, blowing them up even bigger; so we can see some problems that we are likely to face in the near future if we don't change our ways.

In Japan, the average doctor now sees 30 patients an hour. They're called two-minute doctors. You never ask a Japanese doctor a question. The natural authoritarianism of the medical profession has met a perfect match in the authoritarianism of Japanese culture. I had a friend, an American woman, who was in Tokyo. She got sick and went to a Japanese clinic. A nurse took a medical history from her, and she was taken into an office. The doctor came in, looked at her, looked at the history, and wrote a prescription, which he handed her. She asked, "Why do I need to take this?" He didn't answer, but turned red and walked out, leaving her alone in the office. After 15 minutes, she wandered out and asked the nurse what had happened. She was told she had insulted the doctor and that he would not be back. This lack of feedback and dialogue is characteristic of Japanese medicine.

The cost of medical equipment has also had a major effect on Japanese medicine. The Japanese build all of the hardware. They build all of the CAT scanners and MRI scanners. As a result, every little corner hospital has one of these phenomenally

Japanese medicine,
continued

expensive devices. To get the money back, they have to use it all the time. So there's a phenomenal rate of ordering these kinds of diagnostic tests. The rate of coronary heart disease in Japan is a tiny fraction of what it is in America, mostly due to Japanese dietary habits, although they're busily working to "correct" that. But the rate of doing coronary angiograms is the same as in the United States. That means if you have the misfortune to have chest discomfort anywhere in the vicinity of a Japanese hospital, they're going to do an angiogram on you whether you need it or not, because that's the way they recover their money.

The nature of health insurance in Japan has further complicated things. Japan has national health insurance. Everybody is covered, no competitors. Japanese national health insurance reimburses 100 percent for procedures and 0 percent for consultation. So there is no incentive for a doctor to talk to a patient, and there is enormous economic incentive for ordering tests. In Japan now, the national health insurance system is beginning to collapse because it can no longer pay the bills. The same thing is happening in England. The national health service, the pride and joy of the United Kingdom, announced a couple of years ago that it could no longer guarantee access to free medical care for all citizens of the United Kingdom, something nobody ever thought would happen.

The movement away from conventional medicine

Conventional medicine everywhere is experiencing significant problems and economic difficulties. At the same time, there is another vast movement happening all over the world, which I assume many of you are a part of. This movement is a consumer movement away from conventional medicine towards what Westerners call "alternative practice." The statistics are very clear. Now about 40 percent of Americans, and soon will be 50 percent of Americans, use alternative medicine of one sort or another. Most significant, the number of visits to alternative providers last year exceeded visits to primary care physicians. The amount of money spent on those visits exceeded the amount of money being spent on visits to primary care providers. So this is now a very significant market force, and one that can't be ignored. Certainly, at a time when conventional medicine finds itself in such economic trouble, it can't afford to ignore where the market is and where it's moving. And the market is moving very much away from what most physicians are trained to provide.

What patients want

Conventional medicine is simply failing to meet the needs of the patients. In my work with patients I hear the same demands again and again. Patients want doctors who have the time to explain the nature of their problems to them in language they can understand. They want doctors who will not promote just drugs and surgery as the only way of doing things. They want doctors who are at least conversant with nutritional influences on health and who can answer their questions about uses of dietary supplements, which is a huge area of confusion. People say that just walking into a health food store today is totally bewildering. Patients want doctors who are sensitive to mind/body interaction and who are willing to look at

What patients want,
continued

them not just as a physical body, but also as a mental, emotional being and a spiritual entity. They want doctors who aren't going to laugh at them if they bring up questions about Chinese medicine or homeopathy.

I think those are very reasonable requests, but that's not how our medical schools are training physicians. And so, in their frustration, patients are going in greater and greater numbers outside of conventional medicine to other kinds of practitioners. But I feel certain that most people's first choice would be to go to a physician, a medically trained person who is open-minded and knowledgeable enough to guide them through the confusion out there, helping them select the treatments that are most appropriate for them, whether those are within conventional medicine or are a combination of conventional and alternative medicine.

More comprehensive treatment

One example for the potential of this movement is the treatment of Attention Deficit Disorder (ADD) and Attention Deficit Hyperactivity Disorder (ADHD). ADD and ADHD are disorders that could benefit from a more comprehensive approach. They have become very popular to diagnose and are quickly treated with the drug Ritalin. The first question I have about ADD/ADHD is where did this come from? When I was in medical school, this was a very rare condition, and suddenly it became very popular. I always tell people to be wary of fashionable diseases because fashionable diseases have a way of turning out not to be diseases. They go out of fashion. I think it's alarming how readily we make this diagnosis and how readily we prescribe a powerful psychoactive medication to kids. There's no question there's a subset of kids with this diagnosis who respond very well to Ritalin, and it's life-changing for them and their parents. But there is rarely any question of what in the circumstances of a child's life might account for the behavior. There's just this quick-fix method of prescribing a drug.

There are some alternative treatments that have been suggested. There's an osteopathic physician in Texas named Mary Ann Block, who has a center for the non-drug management of ADHD. She uses osteopathic manipulation, homeopathy, and dietary adjustment. There are also a couple of popular books out there. One is *No More Ritalin*. Another is *Beyond Ritalin*. It talks about using homeopathy and dietary adjustment. One interesting area of research I found when writing *Eating Well for Optimum Health* was the role of omega-3 fatty acids in brain and medical function. These are the fatty acids that are in salmon, sardines, walnuts, and flax seeds. One of them, DHA, is the main building block of cell membranes in the brain. There is reason to think that if this is deficient in the diets of pregnant women, nursing infants, and young children, which it is for most Americans, that this may result in disturbed brain function, and one consequence of that could be ADHD. And so a reasonable approach is to try supplementing the diet with omega-3 fatty acids. That's one alternative approach. So at least there's beginning to be some thinking about that.

Terminology: moving towards integration

I rarely use the term "alternative medicine." It's a "button-pushing" term with the wrong connotation. It suggests that you're trying to replace conventional medicine, and that has never been my aim. Obviously, conventional medicine does some things very well, better than any competitors. I often say that if I'm hit by a truck, please don't take me first to a herbalist or a shaman or a chiropractor. I'd like to go to a trauma center and get put back together. But then as soon as I could, I might make use of other methods that I know about to speed up the healing process.

I also don't much like the term "complementary medicine," which first became popular in England and now is in increasing use here, often in the phrase "complementary and alternative medicine," often abbreviated CAM. We now have a National Center for Complementary and Alternative Medicine within the National Institutes of Health (NIH). "Complementary" sounds too weak and "nice." It suggests you're trying to keep conventional medicine as the centerpiece and have little garnishes surrounding the edge of the plate. In fact, when I look at most of what's being done in the name of holistic medicine or integrative medicine, it's mostly complementary. You can put an acupuncturist on the staff of your clinic and toss some lavender oil out in the waiting room and call yourself holistic or integrative, but that's very far from the ideal that I would like to see happen.

Integrative medicine

The term that I much prefer for all of this is "integrative medicine," and I'm happy to see that catching on in academic medicine. There are a number of medical schools, including Jefferson Medical College in Philadelphia, the University of California San Francisco Medical Center, and Duke University, that have recently announced programs in integrative medicine. This seems to be a term that's acceptable in academic discourse. It's non-threatening, it doesn't have the wrong connotations, and it suggests inclusivity. You're bringing things together. I believe this is the term that will be used in the future. However, I often say that the one sign of the success of this movement will be the ability to drop the word "integrative." It will just be good medicine, which is what medicine always should have been.

But how does the integration happen? Who's there to figure out what goes with what? Who uses conventional medicine? Who uses Chinese medicine? Does this herb go with this drug? Who does that? I think that has to be done by physicians, by people who are generalists and not specialists. They should be doctors who have a broad overview of these other therapies, but do not necessarily have to be experts in them themselves. But they do need to know when to refer to them and how to find competent practitioners who know how to put treatment plans together. There's a desperate need for that kind of practitioner, and our medical schools are not training people for it.

A further problem in medicine in this century is that, in its enthusiasm for technology, it has turned its back on all of the simple, low-tech, inexpensive means of intervening in conditions of disease that rely especially on the body's own internal potential for healing. That's both a reason for the expense and a reason for the

Integrative medicine,
continued

dissatisfaction of clients. That has to come back into balance. This movement toward integrative medicine is about trying to restore balance in medicine. As I said earlier, the sign of success is that we can drop the adjective and it will just be good medicine.

Objections and obstacles to integrative medicine

There are tremendous objections from certain quarters in academic medicine to what I'm suggesting. The usual one is that integrative medicine is unscientific, or even anti-scientific. Some people insist that we need to have evidence-based practice and that evidence-based medicine is the only thing that counts. Now, I'm all for that, but there is evidence, and there is evidence. First of all, to keep things in perspective, it's important to realize that a great deal of what's done in conventional medicine is not supported by rigorous scientific study. That applies to many of the treatments that are used in invasive cardiology today and many of the cancer treatments that are being used. These are done because people believe in them. They're done by convention, and the studies are not there. There are many recent examples of techniques that were in widespread use before there was even an attempt to collect the rigorous data. So it's a little unfair to insist that alternative practitioners have to have all this scientific evidence before they go and do things.

Prioritizing research

Furthermore, in an age of dwindling resources, it's unrealistic to think that we can do large-scale, multi-center randomized controlled trials of all of the therapies that are out there that people are using. What we need to do is prioritize those therapies according to which ones have a high potential for harm or for good and which ones are being used by a lot of people and need to be rigorously studied.

As an example, one of the alternative therapies that I would give high priority to is chelation therapy for cardiovascular disease and other chronic, degenerative ailments. This is the process of running intravenous infusions of chelate, an EDTA that grabs onto metal ions such as calcium. The theory is that this removes plaque from arteries. It's an unapproved medical technique that is being widely done in this country. It's probably harmless, but whether it achieves what people claim for it is not at all clear. The little bit of research that's been done on it does not support the claims of chelation therapists. It's expensive and people have to pay out of their pocket for it. Of all the things that are out there, doctors doing chelation therapy is one of the things that boards of medical examiners are most upset about.

Many people are turning to this type of therapy, and there's a lot of testimonials to it from people who claim great results, that they've lost pain in their limbs or chest as a result of doing this, even though the research that's been done absolutely does not support those claims. For a variety of reasons, this is one of the therapies for which there should be a high priority to do a well-designed multi-center trial that will lay to rest once and for all whether chelation therapy is effective or not.

We just don't have the wherewithal to be able to study everything. In the meantime, medical practice goes on, and practitioners and patients must understand that many areas of uncertainty exist.

First, do no harm

If you are faced with a crisis, if a patient is sick, you want to help them, even though the data may not be there to guide your actions. What do you do in that instance? I think you have to rely on first principles. Some of the best first principles were enunciated long ago by Hippocrates. His first one was "First, do no harm." That is a very good guiding principle when you're operating in an area of uncertainty. It's a principle that I think is consistently violated in conventional medicine.

One standard to use is that the higher the potential for harm of the treatment, the stricter the standards of evidence it should be held to. For example, I can't tell you how many women with metastatic breast cancer that I have seen in the past couple of years who have been pushed into having bone marrow transplants even though the research absolutely fails to support the utility of that method. This treatment is very costly, it's painful, and it's risky. There is a significant potential that many women who have this treatment will develop independent cancers later as a result of the high doses of chemotherapy that are used. So if you're using a technique like that, it seems to me it's incumbent on you to be guided by stricter standards of evidence than if you are prescribing an herb that has the potential to upset someone's stomach. That's a different order of magnitude of harm, and there can be a sliding scale of evidence requirements.

Evidence from individual case studies

The evidence that's produced from randomized controlled trials is only one type of information. There are other kinds of information that can guide you as a practitioner working in areas of uncertainty. One is the information that comes from individual case studies, which is often denigrated in academic circles as "anecdotal" evidence. Many people quickly dismiss such evidence, saying, "It's all anecdote." I wish that word would be stricken from the medical vocabulary. It is a biased word. An anecdote is something that an old codger sitting on a porch tells you. It's fantasy. If you want to call this information "uncontrolled clinical observation," that's fine with me, but please don't use the word "anecdote." That's a trivializing word. You can draw conclusions from well-designed, well-collected information about individual cases that can be a guide as to treatments that you use or don't use.

If I hear about a treatment that I did not hear about in my medical training, if somebody tells me they had great success using this or that, my first concern is whether it can hurt people. If I can assure myself that it can't hurt people, then I am more like to pursue it. My next concern is if there is any information published supporting the use of this therapy. Maybe there is, maybe there isn't. Maybe it's research done in other countries or in small populations of people. I also ask if it's plausible. Can I see a mechanism that might explain how this treatment could work that makes sense to me? And then I might want try it out on myself, which is a perfectly valid way of getting information about something. In fact, many physicians in the past felt it was unethical to give a patient a treatment until you had tried it first on yourself. That's an idea that has gone out the window in our modern testing of things.

So even in the absence of data, there are ways of proceeding logically and sensibly

Evidence from individual case studies, continued

when working in areas of uncertainty. This is an important and neglected subject in medical education, because so much of medical practice is areas of uncertainty, and you need to know how to do it best.

Placebo responses

At the same time that some types of valid evidence are denigrated, other types are given too much significance. For some reason, there is an obsession with randomized controlled trials with double-blind studies, but there are inherent problems with that method of data collection. One problem is that one of the reasons for doing them is to separate the placebo responses, which are mind-mediated mechanisms of healing, from intrinsic effects of drugs. But I don't think you can separate placebo responses because to do that would be the same as saying that you can separate mind and body.

One of the strong beliefs I have is that, except verbally, the mind and body are inseparable. I think these are two aspects of the same fundamental reality. I don't think you can ever draw a line and say that the intrinsic effects of treatment are on this side and the indirect effects that are mediated by belief are on the other side. For example, drugs that do well in randomized controlled trials are also going to function better as placebos because doctors believe in the results of randomized controlled trials, and they will present those drugs to patients with their belief.

There's a famous aphorism in medicine that a new remedy should be used as much as possible while it still has the power to heal. An observation that has been made many times in medicine is that drugs work best near the time of their introduction. As time goes by, they work less effectively. Some people would say that this is because with greater experience we see their limitations. But I think a more interesting way and a better way to interpret it is that there is great faith in novelty in our culture. So that when drugs are first introduced, in addition to whatever they do on their own, they are better at eliciting placebo responses. As times goes by, novelty wears off. The halo of placebo response fades, and the drug is left standing on its own, which often doesn't look very impressive.

Maximizing healing responses

There's another aspect of this attempt to separate out placebo responses that is problematic. We devalue placebo responses in conventional medicine. In conventional medicine there's constant talk of ruling out placebo responses. You hear phrases like, "How do you know that's not *just* a placebo response?" The most interesting word in that sentence is "just." To me, the placebo response is what medicine is all about. These are pure healing responses from within that are not mixed up with the direct effects of treatment, which are likely to be caustic. To me, the best medicine is getting the maximum placebo response with the minimum intervention. Not by giving people sugar pills, but by giving them real treatment that you believe in. If you can give treatments that are less harmful rather than more harmful and get maximal healing responses, that's the ideal kind of medicine.

There's an assignment that I give medical students and physicians that I

Maximizing healing responses, continued

recommend for everybody. Go into a medical library and at random pull out any journal in which the results of randomized controlled trails are reported. When you look at the summary table at the end of the article, you will always find a small number of subjects in the control or placebo group who show all of the changes produced in the experimental group. That is astonishing to me. It means that any change that can be produced in the human organism by pharmacological intervention can be mimicked exactly by a mind-mediated mechanism in at least some people some of the time. That includes death, and it includes dramatic recovery from organic physical disease. And whatever that response is, we should be figuring out how to make that happen more of the time. Rather than trying to rule that out, we should be trying to make it happen more often.

Enlarging paradigms

The gist of what I'm talking about is a shift at the conceptual level. We need to look at medicine differently, not just making minor adjustments and adding a few different treatments. I'm talking about a change in our conceptual paradigms of medicine, which demonstrates another limitation that I see in the philosophy of complementary alternative medicine (CAM). CAM is about modality. It's about other tools. It's about teaching doctors to give herbs instead of drugs or in addition to drugs. It's about teaching them to use acupuncture instead of or in addition to surgery. That's fine, but it's a very limited accomplishment. If all we succeed in doing is getting doctors to prescribe herbs, that's useful because it brings in treatments that are dilute and therefore less harmful. It reconnects us with nature. But that seems to be a very tiny bit of the potential there now is for creating change in medicine. The real opportunity right now is for changing things at a conceptual level, for enlarging the paradigm from which medicine operates.

Nonphysical causation of physical events

Let me give you an example of what I mean in the area of mind/body work. Data has been recently published on the usefulness of mindfulness meditation in the management of psoriasis. The reason that this kind of approach is not more enthusiastically embraced in conventional medicine is that it runs up against the paradigm that's in place in Western science and Western medicine. This paradigm is fundamentally materialistic. It says that if you observe a change in a physical system, the cause has to be physical. Looking at mindfulness and psoriasis, or looking at hypnosis as a treatment for ulcerative colitis, or looking at how warts fall of in relation to suggestion, are all examples of nonphysical causation of physical events in the body. That is not allowable in the reigning paradigm. You can't talk about nonphysical causation of physical events.

There are two things you can do with that situation. You can ignore the growing evidence, calling it anomalistic or irrelevant, and refuse to look at it. Or you can use this as an opportunity to enlarge your paradigm so that it accommodates and begins to explain these phenomena of nonphysical causation of physical events.

Focusing on the intrinsic nature of healing

At the moment we have a tremendous opportunity to change the paradigms that now limit our effectiveness as physicians. It would be a great shame if that potential was squandered by simply focusing on adding tools to a doctor's black bag. It's not just about bringing in some of these alternative therapies. It's a chance to learn by being willing to look at things differently.

One aspect of this expanded paradigm is restoring a focus to the intrinsic nature of healing. We need to observe, to teach, to focus on, and to emphasize that healing is an innate potential of the organism. Just as when you get a cut, the cut heals—that potential exists throughout the body. Good medicine should work from the premise that the body can get better and wants to get back to a place of health. Our job as a physician is to help that process along, to identify obstacles to healing, to try to remove them, to give the organism as gentle a push as circumstances allow, in the right direction, and then to let those internal mechanics take over.

An example I often use with physicians is that of a patient who is critically ill with a bacterial infection. You hospitalize that patient, give intravenous antibiotics, and 48 hours later he's out of danger. It is easy to interpret what happens as a cure caused by the antibiotic therapy. I ask them to look at it a little differently. What antibiotics do in that circumstance is reduce the populations of germs to levels where the human immune system can take over and finish a job that it couldn't do because it was overwhelmed. To me, that is a model for how our best treatments work.

Immunization

Another example of utilizing the body's ability to heal is immunization. We're taking advantage of a natural process by facilitating meetings between the immune system and germs at a time in life when the benefits are great and the risks are relatively small. Having said that, I must also mention some reasonable questions about immunizations in recent years. We may have raced ahead and done so many vaccines and done them so early that we have increased the potential for harm. For example, I wonder about hepatitis B and whether that should be a universal immunization. I also wonder about chicken pox, which is a relatively benign disease, but when kids get exposed to it naturally, they have lifelong immunity. We don't know how long the immunity is going to last from the immunization. The consequences of getting chicken pox as an adult are terrible and much worse than getting it as a child. But, although it's not without risks, the benefits of immunization greatly outweigh them.

When treatments work, they often don't work directly. They work indirectly by freeing up, unblocking, and activating internal mechanisms of healing. It's just a different way of interpreting experience, but to me it's much more useful because, first of all, as a practitioner, it's very comforting to know that nature is our ally—that we have this vast force at our back that's waiting to restore the equilibrium of health if we make certain moves or remove certain obstacles. As a patient, that way of looking at things is very reassuring because it's very empowering. It gives patients a sense that there's much that's within their control that they can do to maintain health and restore health. That's a focus that needs to be present in medicine.

Lack of focus on healing

I was astonished in medical school and I continue to be astonished today that healing is not a primary focus of medical education. There should be a whole course in the very beginning of medical school about the nature of healing. This should be a fundamental focus of research. But why, at the National Institutes of Health, isn't there a National Institute of Health and Healing that studies placebo responses, that studies remission from serious diseases, and with a fundamental emphasis on the study of healing phenomena? This is missing at the moment from our medical paradigm. There is now a chance to bring that in, which I think would be a great idea.

Healing centers

Earlier I mentioned that we're going to see the phenomenon of a depopulation of hospitals and clinics due to bankruptcy. We're going to see large areas of the country with one central hospital. The smaller ones aren't going to be there anymore. I have an idea of what I'd like to see take their place. What I envision is a new kind of institution coming into being, which I think we're on the edge of seeing, that I would call a healing center.

This institution would be a hybrid of a clinic and a spa. It would be a place where you could go for three days, five days, seven days. It would not be for the treatment of terminal diseases or critical diseases, but you could go there if you were well, for a lifestyle analysis and adjustment. You could go there if you had any of the common maladies that people have in our culture: anxiety disorders, irritable bowel syndrome, arthritis, back pain, a wide range of problems like that.

These places would be under the direction of generalists, MDs, or DOs, with an integrative perspective, and a variety of other practitioners working under the same roof. In addition, when you came out of there, you'd know more than when you went in about how to eat, how to exercise, how to handle stress, how to use natural remedies, and so forth. Most importantly, staying at these places would be paid for, in whole or in part, by insurance.

I believe doctors would flock to work at these places. They would be places where doctors could attend to their own needs, which they can't now do in conventional medical settings. This is a real possible scenario for the future. And, by the way, there is a model for that in the former Eastern Bloc. There are many countries where there was a tradition of state-sponsored spas. For example, I have a friend whose mother was in Poland and worked for a government factory. She was required twice a year to stay for a week at a spa that was paid for by the government because the government found that workers who did that were happier and more productive. I'm working to help bring some of those models into existence. I think it would be great for our whole society.

Lack of interest in performing studies

One of the obstacles we face in enlarging the paradigm is the reluctance of people and institutions with available resources to fund, support, and perform research in areas outside of the reigning paradigm. This reluctance is just starting to break down.

Lack of interest in performing studies,
continued

Cranial therapy for recurrent ear infections

At the University of Arizona, we have an NIH center, one of about 16 centers in the country. Ours is the National Center for Research in Alternative Medicine and Pediatrics. This center represents a wonderful collaboration between our program and our Department of Pediatrics. We have three core projects. The first one is alternative treatments for recurrent ear infections in kids.

Years ago, in the late 1970s or very early 1980s, I found a wonderful teacher in Tucson who I think is the best clinician I've ever met. I wrote about this incident in *Spontaneous Healing*. He's an old-time osteopathic physician named Bob Fulford who used only manipulation, and was then in his 70s. One of the areas in which he was phenomenally successful was in ending recurrent ear infections in children. I saw this again and again in his office. Working with kids who had ear infection after ear infection, he would do one session of very gentle manipulation on their chests and backs and heads, and they'd never get another ear infection.

The first time I heard the theory from which he operated, I had never heard anything like it as far as medical school was concerned, but it sounded reasonable to me. He said that breathing is the mechanical force that pumps the lymphatic circulation. If children aren't breathing properly, there is stagnant fluid accumulation in the head and neck, including the middle ear, and this provides a breeding ground for bacteria. He said you could wipe out the bacteria all you want with antibiotics, but if you didn't correct the underlying problem, the infection would come back, which is certainly our experience. So he would do one session of manipulation. I watched the kids on the table. Their chests would expand more fully when he finished, and they would never get another ear infection.

I got very excited about this and I thought, "Wow, this is going to be great for pediatricians." So I began calling pediatricians in Tucson and asking them to come and watch. No one would come. I couldn't understand it. To me, the essence of science is open-minded inquiry. If something comes off that doesn't fit your model, you look at it. But that is clearly not the behavior that I see in much of medical science today. It took me several years to get a pediatrician to come down. The one that I finally got was an English woman, who had been trained in England. She came in, watched, and was skeptical, but said she was willing to send some patients to him.

She sent one of a pair of identical twin boys. He was four and had had a very severe ear infection every month. The pattern was that every time an upper tooth erupted, he'd get a severe ear infection. There had been almost no period of his life where he was off antibiotics for more than a month. So Fulford did his stuff on him, and the kid didn't get another ear infection. So after two months, the pediatrician said well, this was interesting, but really not long enough a period. And then, four months went by, and she said, well this was interesting, but this was now the time of year when ear infections were less likely to occur. Well, eventually, after a year went by and he hadn't had an ear infection and upper teeth had come in, she said, "Well, I've never seen anything like this before."

Cranial therapy for recurrent ear infections, continued

That was back in about 1981 or 1982. It is only now, almost 20 years later, that I have been able to get an NIH-funded study of cranial osteopathy to investigate it as a possible treatment for recurrent otitis media (middle ear infections). So when people out there repeatedly ask where the data is, there are real practical problems here. One of them is that the people in this culture who have the money to do research, the facilities to do research, the inclination to do research (because everybody by nature is not a researcher), have not been interested in these areas. Our researchers have not been interested in mind/body interactions or in natural products, for example. Until this situation changes, as a result of changes in education in medicine and science, it is unrealistic to expect the data to be there.

Energy healers

Conventional medicine is very unwilling to look at new things. I mentioned earlier that we bring energy healers in to work with our physicians. This is something that I'm fascinated by. We work with one brilliant, well-known healer from California named Rose LaBriere. She is remarkable at using powers she got spontaneously as a kid. It's interesting that she can't use it on herself even though she's very effective in doing it to other people. We have her come in and work with our fellows for a couple of days to teach them what energy feels like and so forth.

When she was in last year, I had had a cold that had left me with a sore throat that was going on much too long and wouldn't end. So I asked her if she would give me a treatment. She asked me to lie on the table, and she put her hands on me. Instantly, I felt a sensation, as if there were motors in her hands. This was not subtle. There were very intense vibrations coming from her hands. If I would put my hand on her hand, her muscles weren't vibrating. It wasn't something muscular. It was coming off the surface of her hand. Even more interesting, after about five minutes, these vibrations began to spread through my whole body down to my toes. And even more interesting, when she took her hands off they continued for about 10 minutes. She said that some sensitive people would feel this for up to three days after she worked on them. She said she had "amped up my choppers."

Now can you imagine the reaction that I would get from a medical audience talking about somebody "amping up my choppers"? But my experience was something very real, not subtle. It was obvious. I want to know what that is. I can think of various explanations, but I have no idea what it was. And the next day, coincidence or not, my sore throat ended. But it was interesting enough to make me think, well, if that happens again, I'll go back to her, even though I don't have the results of randomized controlled trials to support that strategy.

Calls for evidence

Energy medicine is an example of a field that researchers today are unwilling to look at. Realistically, it will be some time before good studies of it are done. And there is similar lack of research interest in many other fields that are relevant to integrative medicine.

Calls for evidence,
continued

It's great that we have the National Center for Complementary and Alternative Medicine, which is finally funding good work in this area, but is still very insufficient. The budget of this center is less than one-half of one percent of the total budget of NIH. That should be kept in mind. It's also worth noting that this center is not there because NIH thought it would be a good idea to create it. The rest of NIH is bitterly resentful of being stuck with this thing, which is still often called the Office of Astrology. It's there because a group of vocal congresspersons who had control over budgetary allocations at NIH were responding to their constituents who were demanding to know why NIH wasn't doing more in this area that was of such great concern.

The people at the National Center for Complementary and Alternative Medicine have prioritized a lot. They've created a number of these centers around the country, including several centers for the study of botanicals, for example. I think now the emphasis is going to shift onto specific research projects, and they're going to start an intramural research program, which is great.

But I am disappointed that the leaders of the Center for Complementary and Alternative Medicine do not see it as within their domain to support research on medical education for the development of medical curriculum. The position that the director is taking is that the NIH will only fund the training of researchers and will not have anything to do with the training of clinicians or the development of new medical curriculum. The Congress has specifically said to him and to that office that they want to see the development of medical curriculum and training in integrative medicine. It'll be interesting to see how that plays out. Congress is probably going to put very specific language into the budget saying they want to see this office support the development of training in integrative medicine. That's one thing that I would like to see happen.

Medical education

When I talk about changing the four years of medical school, I'm under no illusions about how tough that is to do. For one thing, the medical curriculum is overloaded with information at the moment. Medical faculty already feel absolutely overwhelmed with the amount that they have to teach. They also feel constrained to train and equip students to pass the National Board Examinations. They feel they have no freedom.

Several years ago, I got very upset that the curriculum at the University of Arizona did not include a single lecture on tobacco addiction. Here is the single greatest cause of preventable medical illness, and there was nothing in the curriculum on it. Of course, students heard that smoking causes emphysema and lung cancer, but they heard nothing about tobacco as an addiction and what you could do about that. I proposed giving a one-hour lecture on tobacco addiction and strategies for managing it. It took five years to get approval for the lecture because nobody would give up an hour. The biochemists said they already had too much to teach. How could they give up an hour? If that's the case with a one-hour lecture on something so vital, what chance do you have if you propose putting in a course on

Medical education,
continued

mind/body medicine, or a course on nutritional medicine? That looks dismal.

But the fact that everybody is geared toward getting students to pass National Board Exams gives me an idea. If the National Board Exams started asking questions about nutrition or herbal medicine, faculties would very quickly find it possible to teach about those subjects. There's a way of doing that: by getting a movement among deans of influential medical schools to approach the National Board of Medical Examiners with that suggestion. And that's one of the items on the agenda for our meeting in the fall with a group of deans that we've started to work with who are interested in trying to move medical education in this direction. So I am hopeful.

Consumer-driven change

It's very important for people to realize that the changes coming about in medicine are still primarily consumer-driven. The medical profession and medical institutions are being dragged in this direction, often kicking and screaming. But given the economic reality of medicine today, there is no choice. This is the only direction that they can move in. But however it's coming about, it's a welcome thing. There are now, as I mentioned, a significant number of medical schools that have indicated a willingness to look at transforming undergraduate medical education.

But there is also great resistance out there—some of it because many medical educators feel that changes in medical education should only be guided by advances in scientific research and should not be dictated to by consumers. But what consumers are demanding at this stage is very reasonable. If consumers are demanding that physicians be educated in nutrition, how can you argue with that? If you consider nutrition to be a form of alternative medicine, that's very foolish. Nutrition should be foundational to the development and training of any physician. Similarly, physicians have to have a basic working knowledge of botanicals and of all of these supplements that patients are taking, if only for the simple reason that there might be incompatibilities between prescribed drugs and the botanicals that people are using.

One of the alarming statistics that's come out of the surveys is that the majority of people using alternative practices are not telling their regular doctors that they're doing so. The reasons for that are obvious. People don't want to be laughed at. It's very important for physicians to create a climate in which patients feel trusting enough to discuss what they're doing. It also assumes that physicians have some knowledge of what these things are. I'm always amused when I see newspaper reports of some herbal vitamin that is successful for some condition and in the last paragraph always says not to use it on your own without first discussing it with your physician. What good does it do to discuss it with your physician if your physician knows nothing about it? And there are large areas in which physicians know nothing.

Education attempts

We're trying to correct this at the University of Arizona. The Physician Fellows that we have in a two-year intensive training and the larger number of physicians who are going to start in a nonresidential Internet-based training are the first of a new generation of physicians who are broadly trained to assess the needs of consum-

Education attempts,
continued

ers and patients today. Our aim is not to make the people experts in Chinese medicine, for example, but we want them to know that Chinese medicine exists. We want them to know what its strong points and weak points are, when it's applicable, when it's not applicable, and if you do decide to use it, how to find a competent practitioner of it. Although doctors already have a lot to learn, I see this as part of the basic education of doctors. It's basic information that doctors should be able to help you with. The potential at the moment for making these changes is greater than it ever has been because of the economic climate out there.

Climate for change

It's interesting for me, as somebody who has written and said the same things about medicine and medical education for many years, to see the change in attitude of the medical profession toward what I've been saying. For much of my professional career, most people didn't pay any attention to it at all. In the past couple of years, I've been taken very seriously by colleagues. What's changed in that time? What's changed is economics. And I'm afraid that I have to conclude that the way that institutions change in our culture is not through ideological arguments. Sometimes no amount of philosophizing or ideological arguments will do anything.

But now that the purse strings of medicine are pulled so tight, the institutions are suddenly becoming responsive to these possibilities. The economic forces that are out there are very large. They are not going to go away. The combination of the economic collapse of conventional systems and the expansion of this consumer movement to a mainstream phenomenon has achieved some kind of critical mass. The simultaneous occurrence of these two forces is overwhelming, and medicine really can only move in this direction.

Role of consumers

I want to encourage people as consumers to keep doing whatever they're doing, because clearly that is the force that's propelling these changes. The more that people demand, the more they'll get what they want. Some strategies that I would recommend are to write to their legislators and tell them that they want to see increased funding for the National Center for Complementary and Alternative Medicine. They can write to deans of medical schools and tell them that they want to see doctors trained in nutrition and in alternative medical modalities and so forth.

People can give their own physicians feedback as well. When physicians do things in the way that a patient likes, they should be told. And if there are areas in which they can improve what they do, they should be told that, too. It's very reasonable to suggest to physicians today that they take training in areas that they didn't get in medical school. We're overwhelmed with requests from practitioners who want to get up to speed in areas like botanical medicine and nutritional medicine because they realize that this is important for the practice of good medicine. They realize that this is what patients want, and they realize that they didn't get this training in their medical education. There are mechanisms and ways out there that doctors can take continuing educational courses and other kinds of training that can

Role of consumers,
continued

teach them the basics of all of this. It's very reasonable to hope that, in the not too distant future, undergraduate medical education will begin to change as well.

Conclusion

This is a tumultuous time in medicine. It's a painful time for many institutions and practitioners and patients, but it's also an incredibly optimistic period. This movement has been building in this country and around the world for such a long time, really since the end of the 1960s. It has finally reached the point where we're going to see very concrete results in terms of changes in medical teaching, medical practice, and medical research.

Author information

Andrew Weil, MD, is a world-famous authority on natural healing practices, alternative medicine, drugs and addiction, and the reform of medical education. He is director of the Program in Integrative Medicine at the University of Arizona Medical Center, where he also holds an appointment as Clinical Professor of Medicine. He is the author of seven books, including the international best-sellers Eight Weeks to Optimum Health, Spontaneous Healing, *and* Eating Well for Optimum Health: The Essential Guide to Food, Diet, and Nutrition.

Dr. Weil is a graduate of Harvard College and Harvard Medical School, worked for the National Institute of Mental Health (NIMH), and for 15 years was a research associate in ethnopharmacology at the Harvard Botanical Museum. His web site (www.drweil.com) *is one of the most popular health-related web sites on the Internet. Dr. Weil is also the founder of the National Integrative Medicine Council, a not-for-profit organization in Tucson, Arizona. He has traveled extensively throughout the world collecting information about medicinal plants and healing.*

Editorial assistance for this article was provided by Laurence Smith.

Counteracting the Harmful Effects of Stress through Self-Care to Enhance Wellness and Profitability

Author

Herbert Benson, M.D., President, Mind/Body Medical Institute, and Associate Professor of Medicine, Harvard Medical School, Boston, Massachusetts

Editor's note

Employers and employees are rightly concerned about the quality and cost of health care. Today, affordability is forcing decisions about healthcare coverage options and who pays for it. In this article by Dr. Benson, we see health care as a tripartite system that includes the informed and active participation of the employee and employer to maintain health in the workplace. In the absence of such participation, illness and the cost of health care will continue to escalate out of control.

Dr. Benson describes health care as a three-legged stool with one leg—self-care—being shorter than the other two. The need, to enable good health and to control costs at the same time, is to increase the size of the self-care leg and achieve a stable system.

Experience with large-systems change, however, teaches us that systems tend to be strongly invested in current processes and procedures and tend to resist substantive changes because they are viewed as financially harmful to the system. A system may be populated with many expert and well-intended individuals who see a need and advocate improvement, but within a system, change is slow and difficult to achieve in the absence of a crisis and outside influence. Dr. Benson's pioneering work on stress and self-care—as a practitioner, researcher, and educator—is reflected in this important article—Editor.

Introduction

We at the Mind/Body Medical Institute view health and well-being as a three-legged stool. One leg is pharmaceuticals, the second is surgical procedures, and the third is self-care. Mind/body medicine is the third leg that incorporates all of the following: stress management, exercise, nutrition, belief, and the relaxation response (which I will define later).

The awareness that mind and body interact has important implications for the way we view illness and treat disease. According to the Mind/Body Medical Institute's model, the treatment of illness is most effective if attention is paid to mind/body interactions. Physical symptoms are influenced by thoughts, feelings, and behaviors; conversely, thoughts, feelings, and behaviors are influenced by physical symptoms. Social factors are important as well.

Mind/body science has made enormous progress but has yet to be incorporated as an equal, fully respected partner in Western medical disciplines. As many times as

Introduction, continued

science has affirmed the original message of *The Relaxation Response* over the past two and a half decades, medicine and society have yet to take full advantage of the healing resources within the mind/body realm.

So much has changed in the world: our economy is becoming more globalized, and barriers between countries are being pulled down. But we have yet to witness a corollary paradigm shift in medicine. Today, our appetites have been whetted with quick fixes—so much so that our diagnostic gadgets and miracle drugs have almost overcome common sense. We expect that surgical acumen will be enough to save us, and if not, the next remarkable scientific discovery will.

Although mind/body therapies have been proven effective for the vast majority of everyday medical problems, we are still far more likely to run to our medicine cabinet to relieve our aches and pains than to consider using relaxation or stress-management techniques.

Stress is a fact of life for most working people today. It is estimated that job stress costs employers a staggering $2 billion annually in absenteeism, sub-par performance, tardiness, and worker's compensation claims. Research has shown that more than 50% of adults report high stress every day and that stress is directly linked to numerous medical conditions, such as hypertension, asthma, chronic pain, and allergies, which can account for significant job absenteeism.

Sadly, though, we still rely far more than we should on external fixes—on medications and medical and surgical procedures developed in laboratories—and not enough on our natural potential for self-healing. Therapies we can purchase and caregivers we can consult, whether available through conventional or unconventional medicine, are still far more impressive to us than our own hearts and minds, lungs and hopes, and muscles and beliefs, even though they sustain us day in and day out.

The three-legged stool

The two legs of pharmaceuticals and surgical procedures are absolutely vital to our overall health and well-being. But they take care of only 10% to 40% of the visits patients make to healthcare professionals. The third leg, which I'll be emphasizing the importance of here, is self-care (see Figure 1 on the next page).

Although I will be concentrating on the third leg, it's important to keep in mind that health care must include the other two legs, too. For patients with problems such as heart attacks, broken bones, and punctured lungs, surgery and pharmaceuticals are obviously necessary. No amount of self-care or mind/body practices can deliver what the first two legs of the stool do for these patients.

A definition of self-care

First, let me say what self-care is not. Self-care is not alternative, complementary, or integrated medicine. It differs from alternative medicine for three reasons. First, self-care is scientifically based and evidence based. Alternative medicine is not scientifically based. If it were, it would cease being alternative.

The three-legged stool,
continued

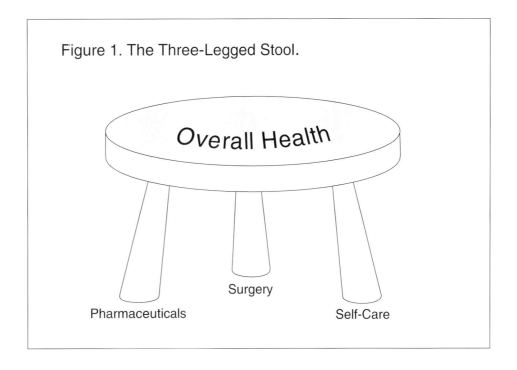

Figure 1. The Three-Legged Stool.

Overall Health

Pharmaceuticals

Surgery

Self-Care

A definition of self-care,
continued

Second, think about what alternative medicine is. There is little difference between administering an herb and administering a pharmaceutical. They are both substances that are given to the patient. Similarly, there is little difference between surgery and acupuncture. They are both things that are done to the patient. The term *self-care* refers to what a patient can do for himself or herself.

Third, when a healthcare provider administers appropriate self-care strategies, patient visits to HMOs decrease by 35% to 50%. Therefore, self-care involves significant cost savings. Compare that to alternative medicine, which is costing the nation scores of billions of dollars per year.

The Yerkes-Dodson Law

Self-care incorporates the principles of the Yerkes-Dodson Law, which was defined at Harvard in 1908. This law stipulates that as stress or anxiety increases, performance and efficiency increase as well (see Figure 2 on the next page). Simply put, when we are under pressure, we perform better. But this trend does not continue upward indefinitely; when the level of stress and anxiety goes up too high, performance and efficiency decline rapidly.

Some degree of stress is absolutely necessary, but too much can cause you to get into burnout situations, as the Yerkes-Dodson Law illustrates. Stress is beneficial if it can lead to production. But the ability to balance it is the key.

Stress can result from any situation that requires a behavioral change. Anything you have to adjust to, be it a deadline, an illness, a family crisis, or a financial problem—situations that companies' human-resources personnel commonly observe among employees—will evoke the fight-or-flight response, which is described on the next page.

The Yerkes-Dodson Law, continued

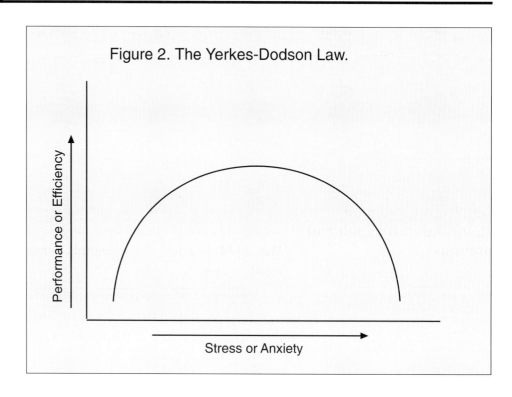

Figure 2. The Yerkes-Dodson Law.

The fight-or-flight response

When you are under stress, you experience an internal pouring forth of hormones from your adrenal glands. The result is an increase in blood pressure, heart rate, breathing rate, metabolism, and the amount of blood flowing to the muscles. This prepares you physically for running or fighting—hence the name "fight-or-flight response."

In our society we don't actually run or fight, of course; it would be socially inappropriate to do so. But that is what your body is being prepared for. It has been scientifically established that this internal injection of adrenaline and noradrenaline (which are catecholamines) leads to an increase in anxiety, depression, anger, hostility, and blood pressure.

Problems related to stress

When you experience stress, your threshold to pain is lowered by the catecholamines that are released. As a result, a vicious cycle is created: when you have a pain and you worry that it might be something serious, you evoke the fight-or-flight response, your pain threshold is lowered, the pain gets worse, and so on.

There are also gender-specific problems directly related to stress. In men, sexual performance and sperm count decrease. In women, PMS, infertility, and hot flashes and other unpleasant symptoms of menopause are made worse both in frequency and severity.

In America, over 60% of the visits made to healthcare professionals are stress related. But stress- and mind/body-related problems are poorly treated by the first two legs of the three-legged stool, our traditional modes of therapy.

Blood pressure readings spawn self-care theories

Now I will provide a background for how I arrived at my theories about self-care. Then I will describe the positive things self-care can do for the overall healthcare system and how it can increase productivity and efficiency.

When I worked as a cardiologist during the late 1960s, I noticed that many of my patients' blood pressure readings taken in my office were unduly high. As a result, I was consistently overmedicating these patients. They would then come back with symptoms of hypotension, or abnormally low blood pressure.

An animal model for stress-induced high blood pressure

I wondered if the stress of having their blood pressure measured could be causing my patients' blood pressure to go up. Together with some associates at Harvard Medical School, I established an animal model for stress-induced high blood pressure.

We worked with monkeys. We measured their blood pressure, fed back their own blood pressure information to them, and rewarded them each time their blood pressure increased. Soon they became hypertensive.

We turned the experiment around midcourse and began rewarding the monkeys to re-enforce lower blood pressure. As we had hypothesized, their blood pressures could then be decreased.

A link with transcendental meditation

At about the same time in California, Robert Keith Wallace and Archie F. Wilson were studying transcendental meditation, and we discovered that our experimental designs where exactly the same. Wallace came east to join me at Harvard, and we collated our data.

We brought young, healthy people into the laboratory and attached equipment to them so we could measure their heart rate, blood pressure, electrocardiogram readings, oxygen consumption, and brain waves.

First we had them sit for an entire hour, and then we conducted the following experiment. It was divided into three periods. During the first period, which we called the premeditation period, they simply sat quietly with their eyes closed and thought regular thoughts. During the second period, the meditation, they sat with their eyes closed and thought meditation-like thoughts. During the third period, the post-meditation period, they thought regular thoughts once again.

Results of the meditation experiments

During the second period, the subjects exhibited a significant decrease—16% to 17%—in oxygen consumption and energy metabolism compared to the first period. There was a corresponding decrease in the amount of carbon dioxide produced, which meant that the decrease in oxygen consumption wasn't through the subjects' increasing or decreasing their respiration. There was a true decrease in the metabolism of the body. There was also a decrease in the rate of breathing, from sixteen breaths per minute to about ten or eleven.

In subsequent experiments working with advanced Buddhist meditators in

Results of the meditation experiments, continued

Tibet, we found that these subjects could lower their metabolism to such an extent that they could almost stop breathing. Their respiratory rates went down to just four to five breaths per minute.

Ruling out sleep and hibernation

During the late sixties when we conducted these experiements, sleep and hibernation were the only two recognized stages during which one could decrease one's energy metabolism from its level at a resting state. We knew that what we observed wasn't sleep. The brain-wave patterns were different. In addition, during sleep there is a slow, progressive decrease of oxygen consumption over three to six hours and then an increase, whereas with the meditators, it dropped within three minutes and stayed at that level as long as they remained in a meditative state. Then it returned to normal.

With regard to hibernation, a good way to differentiate it from sleep is rectal temperature. The rectal temperature of a sleeping animal decreases by one or two degrees, while that of a hibernating animal goes down to almost the freezing point. We measured the rectal temperature of the meditators during the three periods of the experiment, and there was no effective change.

The opposite of the flight-or-fight response

What we were looking at, then, was a physiological reaction that is the opposite of the fight-or-flight response and is induced by transcendental meditation. We named this set of physiological changes the *relaxation response*. But how could it be caused by transcendental meditation alone?

Transcendental meditation has two basic components

We then set out to determine what the components of transcendental meditation are. At first we believed there were four, but now we believe there are just two.

The first component is the repetition of a word, sound, prayer, thought, or phrase, or a repetitive muscular activity. The second is a passive return to the repetition whenever other thoughts intrude.

Religious and secular roots

With my colleagues I searched through the religious and secular literatures of the world to see whether these two steps had ever been described before. We found that every single culture of humankind that has a written history has these two steps described within it.

Normally these steps are described within a religious context. We found examples in Hinduism, Judaism, Christianity, Islam, Shintoism, and many others. In shamanistic religions, people would achieve the same state by chanting in time to the beating of a drum or the stamping of feet.

Outside of religious tradition, we found the same two steps described in the pre-suggestion phase of hypnosis, in autogenic training, in progressive muscular relaxation, and in yoga, tai chi, and chi gong. It is also described in the literature of Wordsworth, Thoreau, Emerson, and Alfred Lord Tennyson.

Experimenting with students

We continued our experiment by bringing some college students into the laboratory and monitoring them in the same fashion as the meditators, attaching equipment to them to measure their heart rate, blood pressure, electrocardiogram readings, oxygen consumption, and brain waves. We measured what was going on physically when they did the two steps and used the number one as a repetition. We found that their physiological changes were virtually indistinguishable from those of the meditators.

Then we brought in another group of people and had them do the two steps and say a repetitive prayer, such as the rosary or a centering prayer. We found that the same physiological changes occurred.

The word, prayer, phrase, or sound that is used doesn't matter. When you carry out these two basic steps, you will experience a measurable, predictable, reproducible set of physiological changes that is the opposite of the set of changes that occurs with the stress response. Under no circumstances should what I am saying here be interpreted as a scientific or mechanistic explanation of prayer. It is simply a reaffirmation of what religious people have told us for millennia: prayer is good for you.

Benefits of the relaxation response

Once our experiments were completed, we hypothesized that the relaxation response could effectively treat disorders to the extent that they are caused or made worse by stress. In other words, those 60% of visits to healthcare professionals that are attributable to stress could be treated effectively.

Furthermore, these patients' performance and efficiency increase; they climb back from the downward slope of the curve of the Yerkes-Dodson Law and start getting back to optimal levels. We published these findings in 1974 in the *Harvard Business Review*, and it rapidly became one of the publication's best-selling issues.

Mind and body are inexorably linked

When one evokes the relaxation response, the result is decreased blood flow across the entire brain. The probability that this is due to chance is less than 10^{-29}. In addition, when a person evokes the relaxation response, the limbic areas of the brain involved with the autonomic nervous system show signs of activity.

In other words, these experiments do away with the concept of mind being separated from body. The brain generates the mind, which in turn affects the brain, as well as the entire body. There is no separation. Mind and body are intermeshed.

Applying the third leg to patients with chronic pain

At a New Hampshire HMO, one of our colleagues, Dr. Margaret Caudill, set up a program to apply the third leg of the three-legged stool to groups of patients suffering from chronic pain. The demographics of the primary site of pain for these patients were as follows: 28% suffered from head pain, 28% had lower back pain, 17% had neck pain, 14% experienced pain in their thoraxes and upper extremities, and the other 13% had pain in the abdomen, pelvic area, and lower extremities.

All of these patients were already utilizing the first two legs of the three-legged

Applying the third leg to patients with chronic pain, continued

stool: pharmaceuticals and surgical procedures. But to that course of treatment we added group sessions.

Makeup of the group sessions

Group leaders included physicians, nurses, psychologists, nutritionists, and exercise physiologists, all working as a team. A physician did not need to be part of every group except when necessary for insurance reasons, and the other leaders could split up the task of leading group sessions. This made the program cost efficient.

Each group was comprised of fifteen to twenty people suffering from some sort of chronic condition. In addition to the general pain-reduction groups mentioned above, our program had groups for cardiovascular disease, insomnia, infertility, PMS, cancer, and AIDS.

The program's five components

The program that each group followed had five components: the relaxation response, nutrition, exercise, stress management (i.e., cognitive restructuring), and the belief system of the patient. Each patient was allowed to choose his or her own focus (a prayer, number, or word to repeat). Together, these five components form the third leg of the three-legged stool: the self-care leg.

We used tapes to teach the patients to evoke the relaxation response for ten to twenty minutes once or twice a day. Immediately after you evoke the relaxation response there is less static in your mind, you listen better, and you learn better. For this reason, we started group sessions with a relaxation-response period and then applied the appropriate nutritional lecture, stress management lecture, and exercise. Normally each group attended about ten weekly sessions.

Results of the chronic-pain program

After the chronic-pain program, we compared the patients' pretreatment status to their post-treatment status. To determine their pretreatment status, we viewed their medical records from the year before they participated in the program. We compared that information to their condition after they attended the group sessions.

Overall, we found a decrease in anxiety, depression, hostility, pain severity, and pain interference (i.e., patients reported that the pain interfered less with their lives). In addition, they needed less social support, their effective stress diminished significantly, and their feelings about life control increased significantly.

Significant cost savings

We had 109 patients enrolled in this behavioral medicine mind/body pain program. One year prior to the program, they were visiting their healthcare facility an average of more than once per month (see Figure 3 on the next page). One year after the program ended, the frequency of their visits decreased to an average of 0.6 per month. This trend persisted for two years.

The program lasted just ten weeks, yet it resulted in significant cost savings. Initially the patients' health care cost $64,000 per year before the program. (These are old statistics; today's costs would be higher.) One year after the program, there

Significant cost savings,
continued

was an overall $12,000 saving even after the cost of the program was factored in. In the second year, there was a $24,000 saving, and this continued.

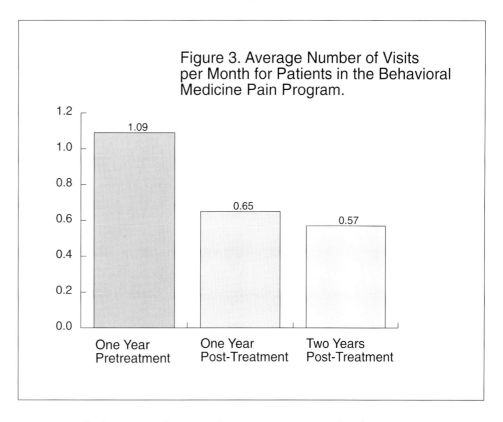

Figure 3. Average Number of Visits per Month for Patients in the Behavioral Medicine Pain Program.

Overall, there was a decrease of 34% in HMO visits for this group. For groups of so-called high utilizers of another HMO, the number of visits decreased by 54% six months after these patients completed the group program.

The program is growing

I am proud to say that we have arranged for the Health Care Finance Administration (HCFA) to pay for our cardiovascular group programs. This four-year demonstration project will involve thousands of patients across the country, and it will be paid for by Medicare.

We at the Beth Israel Deaconess Medical Center have fourteen affiliates throughout the nation who are conducting these kinds of group sessions, in addition to affiliates located throughout the world. Our only problem now is one of maintaining quality and not expanding too quickly even though there is currently a great demand for this program.

How to evoke the relaxation response

Here is the basic technique that we taught to the patients who participated in our mind/body pain program groups. To invoke the relaxation response, first get into a comfortable position. You can do this while sitting cross-legged, kneeling, or even standing and swaying to keep from falling asleep. (If you do this lying down, you are likely to fall asleep.) Then choose a word, sound, phrase, or prayer that conforms to your own belief system.

How to evoke the relaxation response, continued

Close your eyes and relax all your muscles, starting with your feet and working upward. Next, be at ease without moving, and focus on your breathing. Breathe slowly. Each time your breath goes out, say silently to yourself your chosen word, sound, prayer, or phrase. All sorts of other thoughts will come to mind. These are natural, and they should be expected. But when they occur and you become aware of them, don't be upset. Just passively come back to your repetition.

Continue to do this for ten to twenty minutes; then keep your eyes closed, but start thinking regular thoughts. Then slowly open your eyes. The technique is that simple.

We instructed our patients to do this ten to twenty minutes once or twice a day. What is the significance of this? During my research, I looked back to historical precedence of what people did and found that they would pray in the morning and in the evening for ten to twenty minutes at a time.

We then adopted that idea, and we found that it works. More than ten minutes and less than twenty minutes is the optimal time frame. When you are in a stressful circumstance, there is usually very little you can do about it, but if you invoke the relaxation response daily, the likelihood of your reacting to a situation stressfully is decreased.

The relaxation response has diverse applications

One condition that responds well to evoking the relaxation response is hypertension. In our studies, we have seen hypertensive people lower their resting blood pressure, for both the systolic and diastolic readings, by six to eight millimeters of mercury just by evoking the relaxation response. That might not sound like very much, yet over 90% of hypertensive patients have elevations of less than 10 millimeters of mercury.

In addition, 80% of these patients were able to decrease the amount of their medication; of that 80%, 16% gave up their medication completely. These patients all continued to take no medication or less medication over a three- to five-year study period afterward, and their blood pressure remained low. We were disappointed to see that some regained weight they had lost, but their lowered blood pressure was maintained.

We can also help patients effectively prepare themselves for surgery or x-ray procedures simply by teaching them the relaxation response immediately before they undergo the procedure. They don't have to attend any group sessions.

In addition to its applications for better health, the relaxation response can be used to enhance mental and physical performance. For example, athletes now routinely evoke the relaxation response and simultaneously visualize their perfect event over and over again.

The bottom line is that any disorder caused or made worse by stress can be treated by the third leg—the self-care leg—of the three-legged stool.

Getting physicians to accept the third leg

In a survey we conducted of family practitioners, 96% of them said that personal belief and relaxation-response procedures are effective and that they use them in their family practice. When we did the same survey among HMO executives, over 90% of them said that they believed these procedures can heal. However, only 10% said they have applied them in their HMOs.

About two-thirds of physicians today readily accept the concept of the three-legged stool. What often happens with doctors who don't is that their patients hear about these programs from patients of physicians who do. These patients demand that their doctors incorporate them as well.

It's possible to work together to influence insurers

I think people in small and relatively large businesses alike can work together as purchasers of insurance. They can form groups to make their numbers large enough so they will be heard and then refuse to buy any insurance policies from an insurer that does not offer the components of the third leg of the three-legged stool.

Ultimately patients and employers will have to demand use of the third leg. Even though some insurers are balking at it now, they will ultimately agree to pay for these programs because it will be to their ultimate benefit to do so.

Getting corporations to buy in

There is no question that any series of interpersonal relationships—especially those of work—generates stress. We try to do away with stressful relationships, of course, but often you can't do away with all of them. What you can do is have a protective mechanism for dealing with the stress.

The lesson we learn from the Yerkes-Dodson Law is so important. Enabling companies to see that if you decrease stress you increase performance and efficiency will make that lesson more readily adopted.

There is no reason why the relaxation response cannot be applied within a corporate setting. The technique has roots in age-old human behaviors, but it has direct manifestations in our everyday business world.

In addition, we have already taught the relaxation response to industries around the country. Two companies that we work with, State Street Bank and Fidelity, have designated quiet places available in their facilities for people to do the relaxation response on their own.

Enabling people to help themselves

I consistently emphasize all three legs of the three-legged stool. The first two are vitally important, yet it is obvious that they are not doing the whole job. The data I have presented here supports the basis of the third leg. During our work with the groups in the mind/body program, by the third week we consistently hear statements such as "I am a new person," "I am viewing the world differently," "I am no longer bothered by these hang-ups," "It's remarkable that I never thought this way before," and "I am rejuvenated."

We will always need the first two legs of the stool, but the third leg gives us a

Enabling people to help themselves, continued

new approach that enables people to help themselves. I firmly believe it will also help them at the workplace because it makes them happier and healthier, there is less stress, and they become more efficient and more productive.

Author information

Dr. Herbert Benson is an Associate Professor of Medicine at Harvard Medical School, President of the Mind/Body Medical Institute (www.mbmi.org/default.asp), and Chief of the Division of Behavioral Medicine at Boston's Beth Israel Deaconess Medical Center. A graduate of Wesleyan University and Harvard Medical School, he has authored or co-authored over 160 scientific publications and six books, including The Relaxation Response *(1975),* Beyond the Relaxation Response *(1984),* The Mind/Body Effect *(1979),* Your Maximum Mind *(1987),* The Wellness Book *(1992), and* Timeless Healing: The Power and Biology of Belief *(1996).*

Dr. Benson is widely acknowledged as a pioneer in the fields of behavioral medicine, mind/body studies, and spirituality and healing. His work serves as a bridge between medicine and religion, East and West, mind and body, and belief and science.

Editorial assistance for this article was provided by Cathy Kingery and Laurence Smith.

Tools For
Improving the Way
Organizations Run

GOAL/QPC is a nonprofit organization formed in 1979 to improve the economic climate and the quality of work in textile mills in the Lawrence, Massachusetts, area. Teaming with Dr. W. Edwards Deming, we became one of the first organizations in the United States to publicly sponsor his famous quality management seminar. Through our association with Dr. Deming and our continuing research, GOAL/QPC is now a recognized leader in management research, publishing, and training.

How we Improve The Way Organizations Run...
GOAL/QPC offers people practical tools and organizational skills to support teamwork and continuous process improvement within their organizations. Through ongoing research, and development our products and services include:

Creativity	Process Improvement
Facilitation Skills	Project Management
Hoshin Planning	Six Sigma
ISO 9000	Strategic Planning
Leadership	Team Relationships
The Lean Organization	The Journal of Innovative
Organizational Development	Management
Performance Management	Total Quality Management
Problem Solving	

Helping *your*
ORGANIZATION
adapt to a changing marketplace

Our training is supported by cutting-edge research and an ongoing dialogue with the world's leading organizations.

Our Public Courses
are held periodically throughout the United States and are designed to provide a preview of our tools, methods, and processes to the organizational scout.

On-Site Services
can be conducted at your own location or a designated facility. Besides the cost savings, we can also customize the agenda and materials to meet your organization's specific needs.

The Memory Jogger II™ Workshops
The Memory Jogger II™ Trainer Certification
Breakthrough Strategic Planning™
Problem Solving Workshop
Breakthrough Thinking
The Trust Imperative
Project Management Techniques
Facilitation Skills Workshop
Teams Workshop
Performance Management

"I am still reeling (in a great way!) from everything I learned. We have started to implement the tools, and it's been an eye-opener to all!"

Amilyn Lanning
American Color